THE INVERSION FACTOR

THE INVERSION FACTOR

How to Thrive in the IoT Economy

Linda Bernardi, Sanjay Sarma, and Kenneth Traub

The MIT Press
Cambridge, Massachusetts
London, England

First MIT Press paperback edition, 2018

This book was set in Stone Serif by Westchester Publishing Services.

Printed and bound in the United States of America.

Library of Congress Cataloging-in-Publication Data

Names: Bernardi, Linda, author. | Sarma, Sanjay E., 1948– author. | Traub, Kenneth R., author.

Title: The inversion factor : how to thrive in the IoT economy / Linda Bernardi, Sanjay Sarma, and Kenneth Traub.

Description: Cambridge, MA : MIT Press, [2017] | Includes bibliographical references and index.

Identifiers: LCCN 2017019904 | ISBN 9780262037273 (hardcover : alk. paper), 9780262535984 (pb.)

Subjects: LCSH: New products. | Internet of things. | Technological innovations—Economic aspects. | Strategic planning.

Classification: LCC TS170 .B47 2017 | DDC 658.5/75—dc23

LC record available at https://lccn.loc.gov/2017019904

10 9 8 7 6 5 4 3 2

For Ken Traub (1962–2017)

We shared our dream of writing *The Inversion Factor* with Ken, our beloved friend, collaborator, and coauthor. His untimely death, just as our manuscript neared completion, leaves us heartbroken but resolute in our determination to present our (and his) vision of the emerging hyperconnected world, and to enable our readers to participate in shaping that world.

Contents

Introduction

Things Aren't What They Used to Be

History rarely comes with clean lines of demarcation—obvious, self-contained inflection points where one era stops and another era clearly begins. Such moments do occur from time to time: one came to pass in Boston on March 10, 1876, when Alexander Graham Bell spoke into his electromagnetic "gallows phone" the famous words, "Mr. Watson, come here, I want to see you," and his assistant, Thomas Watson, answered. Another such inflection point may have occurred with the founding of Zipcar in Cambridge, Massachusetts, in 2000.

The iPhone was seven years away. Facebook would not debut for four more years. Most Americans accessed the Internet through dial-up modems and Ethernet cables; Wi-Fi was a rarity. AOL and Time Warner were consummating what would become one of the most disastrous corporate mergers in U.S. corporate history. Amazon had yet to turn a profit. Microsoft dominated the tech world by selling office productivity suites and most hardware makers were trying to figure out how they could use the Internet to sell more boxes.

Meanwhile, Antje Danielson and Robin Chase had something else in mind. They knew that in the United States, the average car is parked about 95 percent of the time, translating to millions of potential transport miles sitting idle. Inspired by early Swiss and German car-sharing ventures, they wondered if, instead of developing another vehicle for moving people from point A to point B, they might reimagine how people interacted with vehicles.

The result was Zipcar. The company allowed users to choose from a selection of fuel-efficient cars parked in dedicated spaces around host cities and to reserve a block of drive time. Users would unlock their car with a radiofrequency identification (RFID) card (today, it's a smartphone app) and only pay for the time they drove the vehicle. Zipcar, which Avis Budget Group acquired in 2013 for roughly $500 million, freed students and city dwellers from the need to own a car.

Zipcar also pioneered a range of technologies that foreshadowed the Internet of Things, or IoT: transponders that record time of use and mileage driven and wirelessly upload data to a central computer and a remote "kill" function that disabled the ignition system in case of theft. Zipcar was more than the first start-up in the auto space to use technology to turn shared resources into the "sharing economy"; in many ways, it was the first business to make IoT the cornerstone of its business model. Now, as connected, intelligent, even self-driving vehicles from players as diverse as Tesla and Google begin to move from concepts to reality, it is important to recognize that technology alone did not make Zipcar a pioneer. The x-factor was how they *thought* about the market they were trying to serve.

Rather than ask, "What new technology can we build and sell to people to transport people?" Danielson and Chase chose to ask, "How can we reinvent how people get where they need to go, using technology?" The first question begins the discussion with *product*, and in doing so, automatically limits the answers to variations on existing solutions. By beginning the discussion with *the underlying customer need*, the second question frees entrepreneurs and innovative engineers to unlock fresh solutions to old needs, identify previously hidden markets, create new wealth, and even transform the world's economy.

Things Have Changed

For the history of commerce, things have simply been things. Today, we are seeing a transition from things to systems capable of two earth-shaking behaviors: computing and connecting to other things. When human beings first started using things as tools, function followed form: a rock is big and heavy, so it could be used to break things. A stick is long and pointy, so it could be used to stab things. As humans learned to adapt materials to their own ends, and the craft of manufacturing arose, humanity evolved to a state where form followed function. A designer conceived of some function he wanted a thing to do, devised an appropriate form that would fulfill that function, and then made the thing. If he wanted a thing that would drive small objects, he made a hammer. If she wanted a thing that would cut through stalks of wheat, she made a scythe.

The industrial age gave people an ever-widening palette of materials and construction techniques to use: metals, glass, plastics, ceramics, composites; forging, stamping, milling, welding, molding. Mass production also led to standardized components: screws, nails, beams, wheels, gears, wires, motors, batteries.

Microprocessor technology is the ultimate standardized component, one that can do anything. The development of the microprocessor put limitless power into the hands of the designer. While mechanical functions are limited by the nature of the materials and the laws of physics, in the digital realm, any function the designer can imagine can be realized by the microprocessor. The result: things that can do things that would be impractical—or impossible—by purely mechanical means. You want a microwave oven that lets you set the cooking time to the exact second *and* has presets for popcorn and pizza? Hard to do with a mechanical dial—but no problem with a microprocessor. Digital electronics led to the discipline of user interface engineering, as functions could be finely adapted to the needs of human users.

However, even with microprocessors, a thing is still a thing. It is self-contained and only operates within the confines of its physical shell. It carries out only those functions that its designer envisioned when it was fabricated. The fact that this thing contains a powerful computer inside is a boon to the designer, but it is *completely hidden* from the user. We don't even call this software but *firmware*, reflecting the fact that as far as the user is concerned, it is just as rigid as the hardware. Even if it has a microprocessor inside, a microwave oven is just a microwave oven. We only expect it to do one thing and to do it in isolation.

But because of IoT, our relationship with things—and the very nature of things—has already begun changing. Now the microprocessors inside things are being connected to the Internet through technologies such as Wi-Fi, Bluetooth, and ZigBee (a wireless networking standard ideal for applications such as home automation and medical device data collection) and to each other. A sophisticated network of objects from TV set-top boxes to Internet-enabled baby monitors, each embedded with computers that enable them to collect and exchange data and interoperate, is disrupting business and transforming the way we work, shop, and live.

This new reality—things that are more than things—blurs the boundaries between product and experience. Today, you might wear a Fitbit or other device that monitors steps taken and calories burned. But tomorrow, that same device will morph from a connected activity monitor into the device orchestrating your complete wellness experience. It will reorder a supply of your favorite dietary supplement while syncing with your calendar and sending a text message to your personal trainer to inform her that your Wednesday training session will need to be rescheduled. Based on the intensity of your last few workouts, the device will also talk to the smart machines at your gym and direct them to adjust time or resistance according to your fitness goals. Finally, the same device will detect your elevated post-workout body temperature and communicate with your home's smart hub to preheat the water for an invigorating shower. You won't need to attend to any of these tasks consciously; instead, multiple systems working in concert will combine to meet your needs automatically.

This Changes Everything

The pop culture, *Minority Report*-esque connotations of such scenarios can make it seem as though the changes in progress as a result of the growth of IoT and its associated technologies are novelties. That is an erroneous impression. In truth, the merger of computing, devices, and data is a mega-trend that will fundamentally change business, how we live, and the global economy.

If that seems like exaggeration, consider the current profusion of data and connected devices. The modern automobile rolling off a Detroit assembly line already features hundreds of sensors and multiple computers connected over an internal network that monitors and maintains performance and diagnoses potential system failures. In fact, according to *Auto Blog*, in 2016 Chevrolet alone collected 4,220 terabytes from customers' cars.[1] While most of those connections are local (not connected to the Internet), cars still make up some of the 6.4 billion smart "things" in use worldwide in 2016, according to research by Gartner Inc.[2] According to DHL and Cisco Systems, that number is low: their 2015 trend report, *Internet of Things in Logistics*, sets the current number of connected devices at about fifteen billion, with the total expected to increase to about fifty billion by 2020.[3]

That's fifty billion smart objects—sleek, powerful combinations of sensors (components that convert physical conditions in the external environment into digital information), actuators (devices that make changes to the environment based on digital information), microprocessors (computers), and wireless networking (radio components that send and receive data from remote computers via the Internet) built into objects that a few years ago seemed hopelessly, even charmingly, analog. Do those qualities make such objects inherently more valuable than nonconnected objects? No. As a matter of fact, at this stage of IoT, many "smart" devices are hastily engineered and feature minimal security protocols that make them vulnerable to breaches, and these issues must be addressed. Despite this, the development path of such devices clearly suggests greater utility, increased value, stronger security and privacy protection, and new ways to meet previously unmet needs.

IoT represents the beginning of an exceptionally disruptive era for business. By allowing objects to be sensed and controlled remotely across existing network infrastructure, the foundational systems of IoT allow businesses to integrate the physical world into the world of servers and data analytics, creating an augmented reality of systems where real and digital coexist simultaneously. Products that adapt to user needs in real time, personalized physical environments, smart cities, the potential to discover and exploit previously invisible markets—these are the harbingers of a new economy.

The potential economic impact is significant. Global management consultant McKinsey estimates that IoT has a total potential economic impact of $3.9 trillion to $11.1 trillion a year by 2025, which would be equivalent to about 11 percent of the world economy.[4]

However, this growing ecosystem of smart, hyper-connected things isn't just leading to enticing, seemingly magical new products and services. IoT is powering new ways to meet basic human needs, from transportation to time management. Because of this, it is fundamentally changing business. From smart mattresses that send mattress manufacturers data on customer sleep patterns, to connected valves that allow plant managers to monitor the status of high-pressure water treatment plants, to beacons that will replace traditional traffic signals and use real-time data from autonomous vehicles to control traffic flow, IoT will give businesses an unprecedented cache of data on everything from infrastructure failure patterns to user behavior. By analyzing this data, and by interacting directly with devices in the field, companies can use IoT not only to fine-tune product and service offerings but to reinvent how customers meet their needs.

The "What If" Economy

The business implications of the IoT are not primarily about the Internet or about things. They are about nothing less than a complete transformation of the economy:

> *The global economy is rapidly shifting from an economy of products to a "what if" economy.*

To be fair, many major industries have been characterized by "what if" thinking. For example, before the invention of telephony, an engineer (probably many engineers) had to be thinking, "What if we could develop a way for people to speak over great distances with little or no latency or distortion?" To some degree, all innovators are "what if" thinkers. However, what characterizes our version of the "what if" approach to business is that it is not a means of imagining markets that don't yet exist but of identifying hidden opportunities in markets that *already* exist.

Some businesses already understand this. Rather than organizing their business around the activities of designing products and then packing, shipping, and selling them, these businesses are approaching markets by asking, "What if?"

- "What if people could get products and information at home just by speaking, without even touching a computer?"
- "What if any property owner with a spare room could rent to any traveler looking for a room, and it was as easy (or easier!) to book as a hotel room?"
- "What if you could get a ride in the city from anybody with a car who was willing to take you, just by touching your smartphone?"
- "What if your car could be repaired by a remote software upgrade while you sleep instead of making you go into a dealership?"

We have entered an era when, instead of pushing additional products through a supply chain into a crowded marketplace to satisfy a perceived need, some businesses are saying, "Let's reimagine the best possible way to meet that need." And they are exploiting IoT to bring those visions to life. Instead of shipping boxes with the implied message *This is what our product can do, and all it can do,* they are creating "experiential platforms" that offer those two highly valuable commodities, choice and control: *Here it is; you decide what to do with it and what you'll pay, and tell us how you felt about the experience when you're done.* Choice and control give consumers the power to customize their experience based on convenience, choose from multiple options based on such factors as cost and physical proximity, and to provide immediate feedback to business about what went right and what could be improved. You already know some of the players leading the business charge into the early stages of IoT:

- Amazon
- Google
- Tesla
- Samsung
- Apple

But there are many, many others that you probably have not heard of (yet), including:

- Tile
- Ecovent
- Empatica

- Grab
- Scoot Networks

The list goes on, but the point is that such businesses are no longer outliers. They are disrupting their industries by introducing the versatility, upgradability, and utility of IoT into everything from home security to municipal water system management. And more importantly, they don't constrain themselves by clinging to a line of products. Instead, they focus relentlessly on a consumer need and use whatever technology is required to satisfy it in the best possible way. In some cases, they are creating new markets from scratch by meeting previously unguessed-at customer demands and in a few years, building valuations in that have taken traditional, old-line corporations a century to accumulate.

In the "what if" economy, businesses that survive and thrive will be those whose leaders reimagine the traditional "product first" ideation path of business and instead put needs and experiences first. This concept, which we call *inversion*, is the core of this book. As we will make clear, while IoT is not mandatory to practice inversion, it is one of the most powerful means available for delivering the value that inversion makes possible.[5]

Inversion Defined

The large enterprises of the past, from automakers to builders, were founded on traditional business models that were not designed to respond to rapid change nor to offer their customers choice or flexibility. These legacy businesses were built to do two things:

1. Create products that customers bought and used based on the predetermined palette of features and functions with which designers saw fit to equip them.
2. Periodically modify their products, re-brand them, and sell them again.

This product-first orientation built the world's largest economies and companies, from GM to Procter and Gamble to Nike. But today it is a brake on innovation and growth. Such businesses are restricted to pursuing customer needs that can be satisfied within the boundaries of their *existing product line* or incremental variations thereof.

Before IoT, this product-first strategy could still underpin a viable business because customer data were relatively scarce and the means to gather data were costly and difficult to deploy. Meanwhile, end-user innovation—customers creating their own solutions—was rare. For the most part, only companies whose primary business was moving ones and zeroes—software developers or social networks, for example—had access to relevant data in sufficient volume that the data could drive the evolution of existing products while inspiring new products and services that changed how customers met their needs. And only software companies could spin new iterations of their products on a monthly or even a daily basis.

That is no longer the case. While businesses selling desirable low-tech products (artisanal food providers, bespoke bicycle makers) may continue to thrive while remaining detached from IoT, large corporations in fiercely competitive industries must adapt. This is especially true in sectors where adaptive, tech-forward start-ups are overthrowing centuries-old models by approaching their markets from the same point of origin as Zipcar: seeking not to sell a new product into an old market but to reimagine how an underlying need can be met. Also, thanks to the "maker culture" triggered by entities such as Scratch, Arduino, and MakerBot, end-user innovation has become far more common. We have entered an age of "massively parallel innovation" in which new ideas flow simultaneously from multiple corners of the economy. The consumer's role in the economy is slowly shifting from consumption to collaboration—and eventually, to creation.

For example, Princeton University Professor Edgar Choueiri has developed the first 3D personal sound system, BACCH (band-assembled crosstalk cancellation hierarchy), which can create a 3D personal audio environment for each listener in any space—no earbuds or headphones required.[6] Now, imagine combining your Sonos wireless home audio system, Amazon's Echo, and BACCH, and then syncing them with a wearable device that the Echo would detect as you walked from room to room. The result? A personalized 3D sound "cocoon" that follows you as you move around your home—a completely new way to meet the needs of music lovers through a customized, immersive experience. That's inversion.

But what exactly is inversion?

Inversion is a paradigm in which a business's mission and core competence are defined needs-first, rather than product-first.[7]

Amazon and Airbnb are masters of the inverted business model. Inversion is a strategic philosophy that delivers value not by launching a product and innovating around the product line, but by examining the customer need and innovating around the total experience, using products and technology in whatever way works best to meet that need more effectively and efficiently. A traditional hotel company builds the hotel and might conduct a major renovation every twenty-five years to improve the customer experience. However, since its founding in 2008, Airbnb has completely disrupted the conservative lodging industry by creating a software-driven, user-curated, disintermediated lodging marketplace. In just nine years, the company has accumulated more than two million listings in 190 countries, built a valuation of about $30 billion, and sent hoteliers and municipalities around the globe into a panic.

By taking the focus off of the building and putting it on the travel experience, Airbnb inverted the traditional product-first business model of the hospitality industry and created new and thriving marketplace that it now dominates. That's a clear example of inversion as a core business philosophy.

As Airbnb makes clear, inversion can occur without relying expressly on IoT. Airbnb is not an IoT company at this point it its evolution, but its success story would not have been possible before the development of technologies such as the cloud and the Web—technologies that are fueling both IoT and inversion. While Airbnb and other sharing economy businesses rely on mobile devices to power their business models, connected devices play only a supporting role today.

However, we believe that inversion leads inexorably toward IoT. Even Airbnb is likely to embrace IoT in the future: for example, giving hosts Internet-connected locks to smooth the arrival process for guests. A business with an inverted mission seeks to do whatever is necessary to understand and meet customer needs, and IoT is the technology that lets a business extend its awareness and its influence directly into the customer's experience, in real time. You can even think of IoT as an ecosystem of "convergence devices" that allow technologies such as deep learning/artificial intelligence, cloud computing, wireless/mobile connectivity, and more to be combined for maximum impact. These technologies may not be IoT per se, but they power IoT's highest value capabilities: analyzing massive amounts of data in near real time, seamlessly delivering new services and solutions, recognizing use patterns, and learning user behavior to anticipate needs. So while we will be careful not to conflate IoT and inversion in this book, as we progress beyond the Internet of Things to the more advanced versions of IoT that we will discuss, devices and inversion will become more and more inextricably linked.

Who We Are

After working closely together for almost two decades, the journey of collaborating on this book has been a work of passion for us. Our collaboration started around 2001 when two events coincidentally occurred in parallel several miles apart in Cambridge, Massachusetts. Sanjay was pioneering the technologies and standardization of RFID through the MIT Auto-ID Center (which he cofounded). Meanwhile, Linda and Ken cofounded ConnecTerra Inc., one of the companies that pioneered the integration of RFID into large enterprises.

All this happened at a time when the world had little knowledge of or faith in the applicability of RFID, the Internet of Things (a phrase coined at the Auto-ID Center by Kevin Ashton, one of the cofounders of IoT)[8] was not a known phenomenon, and the world had just experienced the bursting of the dotcom bubble. However, we were confident that the merits of RFID would ultimately lead to its broad adoption, which we championed. Today, RFID is considered to be the genesis of IoT, and we were believers from the start. That's why, over the last sixteen years, we have engaged with Fortune 100 companies around the globe in the understanding and adoption of IoT and continue working at the leading edge today.

Our vision, reflected in this book, is that IoT can help companies meet customer needs and open up new markets to a degree never before experienced. As we continue our work in this field, we are determined to share our vision and insights.

A Roadmap for a World of Connected, Intelligent Things

Inversion changes the object of business from selling products to meeting needs—filling those needs with technology in a way that reduces friction, increases efficiencies, leverages network effects, and turns the customer into a collaborator.

Whether it's a consumer hailing a ride from Lyft, the seamless shopping experience of Amazon's brick-and-mortar Amazon Go, or a hospital seeking remote patient monitoring systems, today's customer is constantly expecting more. By building inversion into its business model, a business can weave data capture, analytics, and new product development into its corporate DNA and with a speed that seems almost precognitive. Consider Amazon's Echo voice-controlled personal assistant. When early adopters brought it into their homes, its use was largely limited to commands such as, "Alexa, please order more toilet paper." But then children starting conversing with Alexa as a friend and the dialogue started changing: "Alexa, what do you think about this movie?" What began as a curiosity became a relationship.

Because of this, Amazon's engineers quickly pivoted, developing systems that leveraged Echo's stream of user data to inspire new products. When a user asks a question of Echo, it goes into a database with an extraordinary level of back-end machine learning. Such deep connectivity between company and consumer generates massive amounts of data that drive the development of new, personalized, and highly intuitive solutions. For example, the Valossa Movie Finder (an Alexa "Skill," Amazon's version of an app for the Echo) leverages the cinema search engine WhatIsMyMovie.com to let Echo users search genres of films (such as mainstream 1990s movies with charismatic leading men) by saying something as simple as, "Alexa, ask Movie Finder to find Tom Cruise movies from the 1990s." That kind of intuitive, seemingly effortless personalization is simply not possible for a product-first company that does not exploit IoT.

To clarify, with its core focus on reinventing commerce and business and meeting needs in new ways, Amazon was an inversion-minded company long before IoT. However, Amazon's vision has come into full focus for the consumer with IoT, primarily in the form of Echo and, to a lesser extent, Dash. For Amazon as for many inverted businesses, IoT serves as the public face of inversion and the means by which value is delivered to the users.

In this book, we have laid out a roadmap to assist the leaders of established businesses in following the footsteps of Amazon, Apple, and companies like them. Consumers who want to know what they can expect from the connected economy and the inverted world can also benefit, getting a preview of what their lives might look like in the future as connected devices become immersive experiences. They might even see how they can participate in this new world as cocreators.

In the first section of the book, we will make the case for why the old, product- and sales-centric way of business is becoming obsolete. We will present evidence to show that the inversion model is an indispensable framework. Finally, we will talk about the need to take advantage of a new vocabulary that business leaders must master to build these technologies and leverage them to serve an inverted mission.

In the second section, we will dive deep into the evolutionary stages of things—the Internet of Things (devices connected to the Internet), the Intelligence of Things (devices that can run software), and the Immersion of Things (devices that interact and cooperate to create seamless, immersive experiences) and how they provide the essential tools to implement the inversion paradigm.

In the final section, we will weave all these insights together into a practical blueprint for business—offering a clear set of steps businesses can follow to make the transition from traditional to inverted—and providing a model for creating an organization founded on the ultimate goal: a culture of inversion.

IoT, Inversion, and How to Read This Book

This book is both an inversion book and an IoT book because each reinforces the other. A company that practices inversion to the fullest quickly finds itself constrained by the boundaries of products that just do one thing. IoT is the technology that breaks these boundaries, allowing the inverted company to become directly involved in the customer experience through remote sensing and to relentlessly and rapidly pursue customer needs by changing what devices do.

Conversely, the full potential of IoT technology can only be realized if a company understands why IoT is transformative from a *business* perspective. Inversion establishes a clear business framework for understanding the potential and use of IoT technology. While it is possible to operate from an inverted company perspective without using IoT (as crowdfunding success stories such as Kickstarter and Indiegogo have done with tremendous success) or to embrace IoT technology under a traditional product-first mission—and we give examples of each—it is the combination of the two that is most powerful.

We have organized the book to show how inversion and IoT work together from both business and technical perspectives. We have mindfully and explicitly developed the content not only to enable all readers to fully become aware of IoT and its possibilities, but also to recognize how to deploy IoT to successfully practice inversion.

Business-oriented readers may wish to spend more time in sections 1 and 3, where we describe what inversion is and how to practice it, but those readers should be sure to at least skim the more technical material in section 2 to understand the stages of IoT technology and how it provides the tools to leverage inversion to its fullest. To break down IoT and inversion into digestible components, we have conducted exhaustive interviews with numerous companies and dispersed these stories throughout the chapters and in greater detail in chapter 8. Technology-minded readers will likely gravitate toward section 2 where we discuss the evolution of IoT in detail. Our presentation is unique in organizing different IoT technologies according to the *user experiences* they enable, rather than typical nuts and bolts of networks, sensors, actuators, and such.

For CEOs, executives, board members, engineers, and product managers, the message is clear. No matter how large your market share or dominant your position in your industry, if you cling to a solely product-first mentality in the age of IoT, you limit your growth opportunities, increase your opportunity cost, cut yourself off from a flow of vital customer data, and hasten your own obsolescence.

For those who aspire to be the CEOs of tomorrow, the message is similar: there are no more siloes. You will no longer be able to say, "I'm a businessperson" or "I'm a technologist." You will either be both or neither.

For the consumer who is simply fascinated by the smart, connected objects showing up in our lives, the message is one of more control, more choice, and more cost savings—but also more threats to privacy. Embrace the exciting potential of the connected, but be educated and aware of the compromises that world may demand.

Disruptions of the status quo make people uncomfortable, provoking uneasy looks and raised eyebrows. This disruption is no different. However, inversion is the key to success in a complex world in which technologies that have developed in parallel must now be integrated. To make that happen, we cannot afford to cling to outdated modes of thinking.

Let us examine the new, inverted world.

Linda Bernardi, M.S.
Sanjay Sarma, Ph.D.
Kenneth Traub, Ph.D.

I Inversion

1 From Products to Needs to Experiences

It is 2030. You get up in the morning with a conference call on your morning's agenda. After breakfast, you walk into your garage and get in your car, where the onboard system immediately recognizes your face as well as the smart device you wear. The car charged itself early this morning, but last night, it actually sold electricity to the grid.[1] You sit down and turn your seat backward for comfort and the car begins your videoconference through the built in immersive display.

Your car drives you to your office, merging seamlessly onto a sensor-studded highway, platooning with other vehicles.[2] There are no traffic lights anymore—just collision-avoiding algorithms.[3] Before taking you to work, your car orders a Starbucks, which you pick up en route, with a caffeine level adjusted for your blood pressure, which your wearable device says is a tad high. A few minutes later, your car books a parking spot and parks itself in an open space for fifteen minutes,[4] transferring your call to your smart glasses so you can disembark and continue the meeting as you walk into your building. Meanwhile, the car knows that it has regularly scheduled maintenance as well as errands to run for you.

At 10:45 a.m., it stops off at a charge station for a battery swap,[5] then goes to the local laundry, where it picks up your suit—delivered by robot.[6] After this, it moonlights for you, delivering packages by enlisting with a "sharing economy" package delivery company, earning some income for you while paying off the carbon debt incurred during its manufacture by running off the clean energy generated by your home solar grid.[7] At 5:30 p.m., it quietly rolls up to the office to pick you up, timing its arrival in front of the foyer as you step out. Meanwhile, a vehicle that you share with another family on your block drives to the local middle school at 3:15 p.m., picks up your kids, and takes them to soccer practice while they do their homework, aided by an artificial intelligence tutor system.

Back in your car, you speak to your virtual assistant and book travel for your family during spring break.[8] The system detects stress in your facial expression,[9] so it plays classical music during the drive home.[10] You walk into your house to the same music, while the temperature has been turned down for your comfort.

This scenario is not science fiction. It might seem like hackneyed futurism, but this technology exists today, albeit in early forms that have yet to be extensively tested. However, in thirteen years, it is quite possible that millions of early adopters living in cities will be following daily routines that look much like the one described here.

Why the Product-First Model Must Evolve

On the surface, the preceding may appear to be a story about technology, and on a superficial level, it is. However, while the technology in use is extraordinarily sophisticated, it is a fundamental change in business thinking—a shift in the forces that drive innovation and the way businesses regard the needs of their customers—that makes that automated morning commute a reality. That is the real story.

To be sure, this transformative approach to business takes visible form as technology. The information infrastructure that will make that car trip and those errands not just possible but *routine*—thousands of intelligent objects and systems interoperating instantaneously with near-perfect reliability—is under development now. Near-ubiquitous wireless connectivity, image recognition algorithms, fast and reliable RFID payment systems, predictive analytics, advanced robotics, immersive home networks, networks of acutely sensitive sensors, computation able to manage terabytes of data, machine learning, artificial intelligence—it is all the product of a thought process that seeks not to move product but to remake how humans interact with the world. It is the direct result of inverted thinking—of a business and technical disruption that is creating systems with evolving intelligence, capable of deep learning, with increasingly seamless integration and interoperability between systems.

One illustrative example is the Nest Learning Thermostat. When it was launched in 2011, it was an innovative take on a familiar home device. Sleek and modern-looking, it redefined the interaction between people and their home. The founders discovered that the average person changes the temperature on their thermostat 1,500 times per year, so they designed the Nest with the capacity to *learn* and anticipate needs, so the person wouldn't have to keep changing the setting. More significantly, the Nest connected to the Internet via home Wi-Fi, so you could now tell your house to warm up as you were driving home from a long vacation or use the temperature sensor to monitor your house while you were away. It was one of the first examples of IoT to reach a large consumer market.

But Nest's use of IoT is only half of the story—the other half is Nest's business mindset. Consider Honeywell, the leading maker of home thermostats at the time of Nest's founding. In 2010, Honeywell's website proudly boasted how they began in 1885 with the first application of a thermostat to control home heating by the Butz Thermo-Electric Regulator Company. This later became the Consolidated Temperature Controlling Company, then the Electric Heat Regulator Company before finally merging with the Honeywell Heating Specialty Company. These company names say it all: Honeywell defined itself by its *product*: heating controls.

Compare this to Nest's mission: "Nest is creating the thoughtful home: A home that takes care of the people inside it and the world around it." They define their mission not by the products they make, but by the needs they satisfy. This is the essence of *inversion*. Nest's product line has expanded since 2011 to include smoke alarms and security cameras. IoT is crucial to this product line, as it enables these products to interact seamlessly to create a total *experience* for the home. Nest leverages IoT to integrate other companies' products, too, and their website lists dozens of other products that integrate with Nest: smart lighting, lawn sprinkler controls, remote door locks, ceiling fans—all products that reinforce Nest's core mission to take care of the people inside the home. Nest has used Internet connectivity not only to break the traditional boundaries of what a thermostat company should make, but also to break through the boundary of its own product line. In this way, IoT is the technological innovation that makes it possible to pursue a needs-based mission without limitation.

As of 2017, Honeywell's website now advertises its new Internet-connected Lyric thermostat by saying: "A home that adapts to you: life's easier with the Lyric family at the heart of your home." It took Nest only five years to force Honeywell to redefine itself after more than a century of defining itself as a heating controls company.

For the first time in history, we have not only the technology, but also the architecture and the new thinking that allows us to anticipate human needs and to meet them in ways that are not dependent on the limitations of the physical world. Advances such as predictive analytics and cost-efficient supercomputing offer customers unprecedented levels of choice and cost savings. But only a relative handful of companies are embracing those technologies and the new mindset necessary to take full advantage of them. For the rest, changing how they think is a matter of survival.

For the many thousands of businesses—many of them small, agile start-ups—embracing the potential of IoT, tens of thousands more remain mired in a product-first business model that either ignores IoT or fails to grasp its true potential. A November 2015 survey by Gartner Inc. revealed that while business leaders are widely aware of IoT, only 29 percent of the companies surveyed were actively using IoT in their daily operations.[11] In the offices and factories of the remaining 71 percent, the process of bringing solutions to market probably looks something like this:

1. Assess the company's existing market for its products or services.
2. Identify unmet or poorly met needs that fit within the company's product or service concept, or opportunities to upsell or replace outdated products or services.
3. Conceptualize and pilot products or services that might satisfy those opportunities.

4. Fabricate, distribute, and sell the products through direct or retail channels; sell the services via existing relationships or communication channels.
5. Hope that adoption and customer satisfaction deliver a positive return on investment.

That process, or a rough approximation of it, reflects the path that new products and services have taken to market for a century or more. To be fair, conventional strategic thinking afforded business owners, engineers, and product managers few alternatives. Whether the end result was laptop computers, soccer balls, motorcycles, or ultrasound machines, the basic paradigm remained unchanged: assess, conceive, design, build, ship, repeat. But in the age of IoT, with our instincts conditioned to the personalization and speed of the tech world, it's not hard to see the deep flaws in that product-first model:

- *It's risky and expensive*. From infrastructure and fabrication to warehousing and service, the costs associated with developing, prototyping, testing, and manufacturing a new physical product—even one that represents a marginal improvement over its predecessor—can be staggering. For vertical industries that depend on sought-after new iterations of its products to be "hits" and produce substantial portions of revenue (automobiles, mobile phones, and Big Pharma come to mind) these up-front costs constitute substantial risk. Even products whose delivery is fundamentally nonmaterial—think Hollywood's summer "tent-pole" movies—require enormous advance expenditures that make them more or less "blockbuster or die" propositions.
- *Interaction ends when the product ships*. When the product leaves the warehouse to ship to the retailer or customer, that's also the end of the manufacturer's relationship with it. The only likely interactions the manufacturer will have with the end user are service calls, complaints, the receipt of warranty registration cards, the occasional market research survey or focus group, and reviews on websites such as Yelp. Hardly the things that rich troves of customer use data are made of—and, in fact, these may be more negative than positive. Some companies conduct extensive market research, but that doesn't provide a real-time, 24/7 flow

of use cases, traffic, reliability, and customer feedback. Consequently, when engineers and designers ask, "What should next year's release look like?," they are often working from fractional, inaccurate data—or no data.

Also, the product-first business is forced to pack everything that it thinks will help the customer into its connected device, knowing that after the product leaves the plant, all the business can do is gather secondhand information about the customer's satisfaction. It can't continue serving the customer's needs or delivering value. Connectivity allows the business to continue interacting with the customer long after the transaction—essentially being alongside the customer offering additional functionality, responding to requests, and providing new value as the technology evolves.

- *It's rigid*. Function remains bound to form. Even a home audio system with a microprocessor in it is still just a home audio system unless it does the thing that defines IoT: connect to the Internet. Teams design products with a single function in mind, and the customer is limited to that function; there is usually little or no ability for the customer to say, "I want to make this do B instead of A." Even services, which are by nature more fluid than physical products, tend to be static and rigid because so much is invested in developing and selling them into their target markets that they adapt to customer demands reluctantly, if at all.
- *It's unresponsive*. In product-first businesses, upgrades usually arrive at a snail's pace because the new product has to be designed, a few new bells and whistles added to justify the cost of upgrading, manufactured, and shipped. For some companies, the rollout of a new version is an annual ritual. On the other hand, it's not uncommon for software as a service businesses (companies that sell subscriptions to software that's hosted on a remote server, so their customers own licenses to use the software, not the software itself) to upload bug fixes or updates of their products weekly or even daily, sometimes overnight while the customer is sleeping. Such companies are continually interacting with customers and gathering data; that is part of their business model. This allows their products to evolve in response to changes in the marketplace. Contrast that with a timeframe of about three years—from customer profiling

to design, rough build, prototyping, tooling, and manufacturing—for a new car model to roll off the line. The slow mover is at a distinct disadvantage. By the time they release a laboriously crafted new version of something, the entire market could be unrecognizable.

• *It's brand-limited.* In 1982, Colgate (before it became Colgate-Palmolive) decided to launch a line of frozen meals called Colgate Kitchen Entrees. The company saturated the airwaves with TV advertising at up to $16 per impression and went all-in with coupons, trying to grab market share from Lean Cuisine. The product bombed. In the minds of consumers, Colgate was a toothpaste brand, not a food brand.

*When you focus on products, those products **become** your brand.*

Consumers strongly associate companies with the category of products they produce, and it becomes extremely difficult to see them as anything else. When product-first companies try to venture outside their core product category, even if the result is brilliant, the odds of failure are high.

A company such as Google doesn't face that same obstacle. Google may have begun as a search company, but it has become far more, an innovator in such diverse areas as advertising, broadband connectivity, life sciences, and through acquisitions such as Nest and YouTube, home environmental control and consumer video. Because it is not linked to any one product or product category, Google faces no perceptual headwinds when it ventures into new areas such as space flight. In fact, bold innovations in general, not any particular product offering, define Google's brand.

Above all, the product-first model is vulnerable to theft, sabotage, and competitive pressures because the businesses that adhere to it are slow to adapt to disruption—and disruption of the status quo is the philosophical operating system behind IoT and inversion.

Taxi!

An excellent example of this is the wave of change affecting the taxi industry. Horse-drawn hackney carriages for hire started operating on the streets of London and Paris in the seventeenth century. While the vehicle itself was replaced in the early nineteenth century by the agile, fast hansom cab and finally by the gasoline-powered taxi, the business model remained unchanged. A fleet of vehicles for hire, licensed and regulated by a governmental agency, delivered passengers to their destinations for fixed fares. The major technological advances of the taxi industry—the taximeter, which calculated and displayed fares based on travel time and distance, and the radio dispatch system used to assign a cab to pick up a passenger at a specific location—didn't change the business model or the customer experience.

While the business environment for taxis has seesawed between deregulated and protectionist since the time of England's King Charles I, it hasn't made much difference to the people who ride the taxis. Finding a taxi has always been fraught with anxiety, whether you are trying to hail a cab on the street or waiting on hold for fifteen minutes to speak to a dispatcher. In many cities (London is a notable exception thanks to its splendid cabbies and their mastery of the dizzying collection of urban data points known as The Knowledge), it's a crapshoot whether your driver will know where your destination is and how to get there. And the vehicle itself might be clean and safe, or it might be horrific. About the only thing you can count on is that your driver, who may be professional but may also be obnoxious, will expect a large tip at the end of the ride.

Then came Uber, sending taxi companies into a panic not by *competing* with them but by *reimagining* the experience of moving people from point A to point B. Frustrated with rude drivers and filthy cabs? Here's a passenger rating system that holds drivers accountable. Tired of getting blindsided by a high fare? This app lets you know what your fare will be and charges your credit card; you just climb in and hop out. Hate the monopolistic practices and cronyism of the medallion system? Here's an upstart alternative designed for *your* convenience.

Even more profound than Uber's reimagining of the taxi experience is how it defines its mission. Uber doesn't define itself as a taxi company; its mission states: "Whether it's a ride, a sandwich, or a package, we use technology to give people what they want, when they want it." Over time, Uber augmented their "black car" offering to include uberPOOL (carpooling on demand), uberLUX (luxury limousines), uberRUSH (package delivery), uberEATS (food delivery), and more. Uber is not simply a better taxi company, *it is in a completely different business*. That Uber competes with taxis is a side effect of Uber's reimagined mission, not its primary goal.[12]

Uber and ride-sharing fellows such as Lyft have disrupted the taxi market by changing their users' relationship with the physical world. In doing so, they have had profound economic effects on the taxi markets in many cities. For example, according to the *Los Angeles Times*, from 2012 (when Uber and Lyft began operating in Los Angeles) to 2015, trips by conventional taxi fell by nearly 30 percent, from 8.4 million in 2012 to a little more than six million. A spokesman for LA Checker Cab says that take-home pay for conventional cab drivers declined by as much as 50 percent over the same period. Similar effects have appeared in other cities such as Las Vegas, a sure sign of disruption of a long-standing business model by aggressive upstarts.

A Core Competence in Experiences, Not Products

This story illustrates the perils of a product-oriented *core competence*. A core competence is that by which a business defines its mission and distinguishes itself from competitors. For hundreds of years, the core competence of most businesses has been designing, manufacturing, and selling their physical products or services. A product company's primary focus will be on its product or line of related products. To sell those products, the company might also develop "solutions," each of which aims to define a specific context into which its products can be sold.

General Motors makes cars. One market context for GM cars is sales to individual consumers; GM's solution for that market includes their dealer networks and service centers. A second context for GM cars is fleet sales to rental car companies, police departments, and so forth, each of which requires a unique sales force, marketing, and other supporting functions, all different from the consumer dealer network. However, the products themselves remain largely unchanged across the different solutions, except for some small customizations like a heavy-duty suspension for law enforcement use.

In a *product-first* company, engineering's primary purpose is to ensure that the product meets its specifications at the time it is shipped. Companies employ rigorous design, manufacturing, and quality assurance practices to this end, but customization and even innovating may be regarded as cosmetic—ways to repackage the same products to resell them into a variety of vertical markets. The emphasis in the product-first model is not on solving new problems or meeting new needs but on keeping the pipeline full of products and marketing messages for the new customers that are the company's growth engine.

For years, this model served businesses well, created trillions in wealth, and became the bulwark of the world's strongest economies. However, factors such as corporate debt, ego, arrogance, and shareholder pressures are blinding product-first businesses to the dangers embodied by bold corporate thinkers and entrepreneurs who are embracing IoT.

Meanwhile, agile businesses can look through and exploit a different *frame*—a distinct way of viewing their products, missions, and goals. Great companies such as Amazon never hold a static view of what their business *is*, only what it *can be*. They understand that the frame must change to suit the customer. These are not businesses that play by the old rules. Their core competency is not building and shipping units, but in *meeting a specific need and delivering a certain experience*. By exploiting all possible means to meet needs and create experiences, including and especially IoT technology, such companies put choice, control, and feedback into the hands of the customer and offer value with which product-first businesses simply cannot compete.

A product-first business that falls in love with its own products and ignores underlying needs and the desire for experiences—a state we call "frame inertia," the inability to define a product or company as being anything more than it already is—can become blind and slide into irrelevance with shocking rapidity. Consider Motorola and its Razr mobile phone. Motorola invented not only cellular telephony but its infrastructure, and its sleek, metallic Razr V3 flip phone, introduced in 2003, became the colossus of the industry, selling more than fifty million units by mid-2006.[13] But without knowing it, in 2006 Motorola was poised for decline. In September of that same year, the Canadian company Research in Motion (RIM) released a bold new redesign of its BlackBerry device, the 8100, otherwise known as the Pearl. The Pearl became the go-to mobile communications device for businesspeople, entrepreneurs, and professionals.

Years earlier, RIM had changed the game with a new vision of how people would use their mobile devices: not just to talk, but to communicate by e-mail. The new BlackBerry crushed the V3 not because it offered just a refreshed product but a new *experience*. RIM reconceived its core competence not as making phones, but of *helping people communicate on the move*, whether by voice or e-mail. Motorola was locked into a business model based on rebranding new versions of the same old phone. By the fourth quarter of 2007, the Razr brand had faded into obsolescence.

Ironically, Apple's 2007 debut of the visionary iPhone, based on an even more expansive mission of changing how people experienced their entire world (with the smartphone, which is IoT-in-a-box, as a sort of IoT magic wand), set off RIM's own period of desperate countermoves and an eventual slide into irrelevance—a cautionary tale and reminder that product-first, "What can we sell them?" thinking is an ever present peril.

IoT and the Maker, Hacker-First Mentality

We would be remiss if we did not call to attention one of the other important factors provoking change in the product-first model: *maker/hacker culture*. The explosion of rapid prototyping tools available to a creative person and a new mentality of iterative innovation and development is changing the way that products are developed.

Take rapid prototyping. Three-dimensional printing (which is not the whole story but a useful way to explore this topic) shows us how today's designers can find instant gratification when they have a creative thought. Just print the object and see how it feels or works, then change it and iterate. A range of people from hobbyists to serious designers are using 3D printing and other mechanical fabrication technologies—desktop computer numerically controlled milling machines or laser cutters, for example—to rapidly bring ideas to fruition. This trend launched the modern maker movement, reviving interest in hardware. Large companies beware—crowd creativity has been unleashed, and we are in the era of massively parallel innovation, where consumers are modifying existing products and even creating their own.

Rapid prototyping can also be applied to electronics and software, where there have been similar advances in tools that shorten the time from idea to action. The Arduino is a family of easy-to-program, open-source microcomputer development boards that come with add-ons that have revolutionized the learning, teaching, and making of DIY electronics—and consequently, IoT. The Raspberry Pi is a small computer running Linux that can be purchased for less than $40 that takes the power of DIY electronics to a new level. Developers can add Wi-Fi, Bluetooth, ZigBee, cellular, and more to the Arduino and the Raspberry Pi; it is not uncommon to see middle school and high school students at maker fairs showcasing work that large companies would (and should) envy.

A revolution in DIY software productivity is happening in parallel. Programming languages such as Python, frameworks such as Django, and environments such as node.js have made software development, both in the cloud and at the edge (the computers at the outer reaches of a network, closest to the end user, as opposed to the servers at its center), accessible, powerful, and quick. Young people can now teach themselves how to program on YouTube and get up to speed in a matter of days. Even advanced concepts such as machine learning have become accessible because of the availability of open-source implementations such as TensorFlow, an open-source machine learning library used to build and train neural networks to identify patterns.

Iterating to the Sweet Spot

The story that's not getting much attention is how these makers are turning their ideas into prototypes and then money. No longer are they limited by budget review committees and stage-gate processes. Rather, the focus is on agility and, to quote Eric Ries, "maximizing the number of iterative cycles per dollar."[14] Have an idea? Don't waste time—just make it.

A consummate maker might turn an idea such as a new electric skateboard into a prototype within a week. Software cobbled together from GitHub, motors from Alibaba, batteries from RC Hobby Pro, and other parts from McMaster-Carr. Some parts are 3D printed and others laser cut or water-jetted. Bingo—a working prototype. Perhaps you want to track your rides: GPS module, no problem, build in a modem, and leverage the cloud. Want voice recognition or text to speech? Use a cloud-based service. Easy.

Next, test it. See if it works. Forget confidentiality; speed of execution and your passion will trump the copycat. Show it to your friends. Get feedback and turn it around. Test and repeat. If it works, you don't need to get shelf space at Best Buy or Target. You *pretail* it, creating interest while you raise money on "need sensing" platforms such as Kickstarter and Indiegogo—gauging demand for what you've made. If enough people contribute or preorder, your experiment has succeeded. If no one buys, you've gathered valuable data to take back to the drawing board.

Rapid prototyping and crowd funding are not simply fabrication and revenue generation methods. They are tools for changing how a business does business. They can bring ideas to life with incredible speed, shaping and evolving them to meet customer needs and generate revenues while large corporations might still be conducting focus groups.

The Solution Mentality

As time-to-market shrinks and IoT insinuates itself into virtually every corner of human enterprise, from smart forks and toothbrushes to self-monitoring oil drilling rigs, more and more businesses are beginning to see the wisdom of freeing themselves of the limitations of product-first strategies. Small start-ups have an inherent advantage in this. They lack the size and resources to compete with a Sony, GE, or Johnson & Johnson, so they don't try. They have no choice but to look for ways to identify hidden needs, unlock value, create new markets, and place big bets with big upsides. They are like value investors looking at underrated companies on the stock market, trying to find what others have missed.

When such entrepreneurs find an unmet need that they can monetize, they have many tools at their disposal. Not only IoT, but also mobile devices and machine learning can be used to offer the customer a customized, low-friction experience. If this experience is sufficiently novel and appealing, and creates value in a way that is meaningful to their customers, they can earn not only those customers' business but their allegiance. Most such ventures fail, but those that succeed can do so impressively, birthing lucrative new markets and disrupting entire legacy industries.

Here, IoT becomes an essential enabler of inversion. For companies operating in the sharing economy, a mobile phone and a software app can be the foundation of a successful business. However, beyond that economy, a connected device that gathers data, analyzes it, and gives the user actionable information can be the entry point into a market where meeting needs supersedes the importance of moving product. For example, a Fitbit that informs its user that she has not yet met her daily step goal is recruiting its user's mobile phone connectivity to fully leverage the value of its data. The IoT device is essential to the comprehensive solution.

However, large businesses can follow this same solution-centric path even if they are not active players in IoT. Google is one of the world's most valuable corporations, yet it excels at rolling out customer-driven new products and online services, solution-oriented selling, and keeping in close contact with how customers use their products. In part, this is because they have been deploying machine learning (computer programs that can change operations when exposed to new data) and deep learning (a class of machine learning that employs neural networks and is a foundation of artificial intelligence) for more than a decade. Google may have more than 60,000 employees, yet it can operate with the agility and *elan* of a start-up because its sophisticated approach to customer interactivity, testing, and data gathering has been baked into its business model from the beginning.

Amazon Dash

When Amazon introduced the Dash—a Wi-Fi enabled button with a product brand name emblazoned on it (Huggies diapers, for example) that automatically orders the product when pressed—many thought it was an April Fool's joke. It was however, an extraordinary act of inversion. What made it so was that Dash followed the rollout of Amazon Prime, which replaces the incremental cost of shipping with a one-time fixed cost—effectively unbundling the shopping list. Now it became possible to order items as needed, and the Amazon Dash brought "one click shopping" to the physical world.

Far from being a practical joke, the Dash was Amazon's deadly serious step toward the hyper-connected world. What if Dash comes pre-installed with your next Whirlpool washing machine? Just as you might get an iPhone for a discount when you're under contract with Verizon, perhaps your Whirlpool will come with a three-year Tide replenishment contract with Amazon. Perhaps we will think of "washing as a service" instead of washing machines.

A classic example of solution-orientedness is Amazon and its experiences with the Kindle. A product-oriented company would have declared victory after the Kindle achieved dominant market share among e-readers. But Amazon's goal was to reshape and own the reading experience, so it released a Kindle App for smartphones and tablets, thus compromising its own product but winning the market. This wasn't an IoT-enabled change, but it did reflect the fast-fail, fast-pivot mindset needed for the IoT world. It was an excellent example of prioritizing a need over a product—inverted thinking.

This approach is possible for a large legacy company, but it comes with obstacles. The greatest challenge for an established organization is to take the leap away from product orientation without pulling resources away from the core competences that pay the bills. But when they find that they can sate a deep, personal need or give people a solution to a problem they believed unsolvable, they can create big wins at a scale that small companies can't match.

For example, according to a Harris Poll survey commissioned by Chevrolet, parents with teens worry about their child driving without supervision (55 percent) more than any other area of parental stress, including drinking and sex. In response, Verizon created the Hum, a connected device that plugs into your car's on-board diagnostic port. The device can contact roadside assistance for you or translate diagnostic messages, but the Hum's sweet spot is that it lets parents set up speed and boundary alerts so they can monitor their teens' driving habits via the cloud using their mobile devices.

The new business thinking embodied in IoT has the potential to transform business in start-ups with a single-digit headcount as well as corporations trading on the S&P 500. However, to fully leverage the opportunities of this era, business leaders, entrepreneurs, engineers, designers, developers, and product managers must understand how fundamental business concepts are changing, powered in part by IoT's ability to put business in close proximity to the end user, where they can gather user data with unprecedented immediacy and relevance.

Six Business Fundamentals That Are Changing

This section looks at how six foundational business ideas have long been approached in the world of product-first business strategy. These are not points of failure, but they are important ideas that IoT is changing—sometimes dramatically. In chapter 2, we will see how new approaches to each of these are required, along with new fundamentals.

1. Brand

Brand has long been the Holy Grail, particularly for businesses selling mass market consumer products. Traditionally, a brand can be thought of as the collected qualities of a product or service and the promise they communicate to the buyer. Brand is regarded as the linchpin of corporate value, to be safeguarded at all costs. In many ways, the brand *is* the traditional, pre-IoT product because in the consumer realm companies are as often selling a feeling of belonging, style, or "cool" as they are a physical object with beneficial properties.

In many cases, the company brand surpasses the product brand and even lends products a "brand halo," a feeling of being inherently superior by being associated with the company. An obvious example would be Porsche, an automotive brand whose name and cars, with their distinctive look, feel, and performance, inspire not only loyalty and love, but also a fierce sense of belonging among its drivers (if you've ever seen anyone give the "Porsche wave," you understand what we mean).

In the product-first world, brand is closely linked not only with the physical traits and appearance of the product, such as its design and materials, but also to the social and cultural cachet created by marketing. Hence the billions spent each year on brand marketing campaigns to imbue products with ineffable qualities.

2. Competition

Competition has long been a simple, Darwinian proposition: every man for himself, and every company in the same sector is an enemy, an acquirer, or a target for acquisition. Companies have clear-cut boundaries where they end and others in the same vertical market begin, and barring a formal partnership or strategic alliance, they work to seize market share from one another. Change happens slowly, at the pace of product rollouts and incremental improvements. There is room for numerous players of various sizes in many markets because customers have limited access to information allowing them to compare the different providers' weaknesses and strengths.

Perhaps most importantly, competitors rarely change (because companies that are dominant in a specific industry tend to limit themselves to that industry), and collaboration is always intentional, never occurring by accident.

3. Innovation

Innovation is about reimagining not only the solutions to problems but the problems themselves. However, true innovation is rare. Instead, most of the innovation in a product-first model is in name only—subtle, incremental, often surface-level tweaks in existing products. When something takes the world by storm, such as the Sony Walkman or iPod, it's usually not because it represents anything genuinely novel but because its blend of form factor, marketing, and timing hit the market at just the right moment to be successful.

Innovation in services is even rarer, quite simply because there are few truly new human needs that can be identified and met by the tools of the product-first world. Even such innovation as does exist—the field of customer relationship management taken to the mainstream by Salesforce, for example—is more a tech-enabled variation on an old theme than a true innovation.

4. Market Share

Market share is the primary metric of success in an environment where strict boundaries exist between companies in the same market sector. Each company's main goal is to expand its market share; when expanded, its goal is to defend that beachhead at all costs, against all enemies, foreign and domestic.

One of the most critical aspects of market share is that a business can grow its market share in one of two ways: increasing the size of its market or taking market share from competitors. Most often, in the product-first world, businesses seek to capture market share from competitors because the opportunity cost of expanding their market—for example, opening a network of dealerships to sell cars in developing countries—is simply too great and the barriers too high to be practical.

5. Ownership

This is one of the clearest concepts in the product-first world. Because the boundaries between organizations are clearly defined and impermeable, so too are the products developed and distributed by each organization its sole property. Services differ only in that they are nonphysical; a web of intellectual property laws protects processes, creative ideas, designs, corporate identities, and even distinctive retail and restaurant designs and customer experiences. Meanwhile, trademark and patent laws drawn clean lines around the physical objects each business produces.

6. Value

In this ecosystem of fixed boundaries, little agility, and low adaptability, value is largely about what a product or a service does. For a motorcycle, value equals how it performs and handles and its reliability. For a motorcycle helmet, value equals comfort, beauty, and durability. With a motorcycle repair shop, value equals the skill of the mechanic and the promptness of the repair. For the network of motorcycle dealers, value equals the frequency and sincerity of its communication with customers and the promptness of its action on issues such as recalls.

Notable about this measure of value is the fact that companies rarely, if ever, deliver unexpected value beyond the limits of the product or service. Motorcycle helmets protect the face and eyes from wind and road debris and provide protection for the head in the event of an accident; they do nothing else, and no one expects them to. On occasion, a desirable brand delivers value in the form of social status or an ineffable "cool factor," but that type of value is fleeting and capricious.

A World Defined by Vision

The true limitation here is not technology or expertise, but *thinking*. Technology, especially IoT technology, is making it possible to imagine services and markets that were not conceivable a few years ago. Even as recently as a decade ago, mobile network data speeds, smartphone market penetration, and GPS accuracy would not have allowed ride-hailing companies such as Lyft, Didi Chuxing, Ola, Grab, or Uber to exist in the fast, convenient way that has made them one of the bellwether categories of the sharing economy and some of the most disruptive companies in recent memory. Many revolutionary ideas perished on the vine because the technology to make them a reality did not exist.

The material and digital things that make up IoT—small, light, cheap sensors that gather massive amounts of data and actuators that interact with the physical world, along with the machine learning that helps Amazon roll out new products and operates Google's prototype self-driving cars—are relatively recent inventions that have permanently changed the game. (Interestingly, self-driving cars are technologically at the midpoint between car design and autonomy/mapping technology. It's revealing that Google thought it easier to design a car than GM or Volkswagen thought it to develop self-driving technology.) In turn, they have been triggered in large part by competition in the smartphone industry—smartphones being, as we will see, the conductor's batons of IoT. However, all these tools remain just tools, nothing more. The real game changer has been the vision they have empowered.

All solutions, disruptions, and paradigm shifts begin with smart people asking better questions, caring about the answers, and envisioning a way that one part of the world can operate in a different, better fashion. In the past, the product-first philosophy limited our imaginations to the factories we could build, the energy sources we could harness, and the objects we could fabricate. No more. We are transitioning from a *technology-defined* world to a *vision-defined* world.

In the IoT era, what a business can achieve depends more on visionary thinking—on "What if the market for that product or service worked this way?"—than on the availability of technology. No matter how audacious the idea, the odds are good that the solution, from natural language processing to machine intelligence to robotics, is already available or on the development fast track in places such as MIT, Google's X research center, Carnegie Mellon University, Technion, or Shenzen. For the first time, whatever entrepreneurs can imagine can probably be done.

For centuries, new markets have arisen slowly as products and customer needs evolved. One hundred years ago, if a business moved into a new market, ten or twenty competitors would be making the same move. Even in the last few decades, creating a new market for a product meant spending millions on advertising, marketing, and brand development. If a company launched something truly novel—digital presses to replace printers' old offset presses—they had to shoulder their way past heavy skepticism.

Today, it is possible to tap into a resource that has traditionally not even been viewed as a marketplace or center of economic activity and "marketize" it. The sharing economy, from Lyft to TaskRabbit, SnapGoods, and Roadie, is but one example. It came to pass because entrepreneurs perceived hidden economic value in peer-to-peer collaboration empowered by fast, intuitive, connected things. Companies no longer need to fend off competitors all trying to sell a different variation on the same product. With vision, insight, engineering and design skill, and a passion for disruption, anyone can be first. Those that do it will embrace a completely new business model: *inversion.*

2 Inversion

When Michael Keating and Dan Riegel decided in 2015 to bring the speed, convenience, and low carbon footprint of the electric scooter to the commuters of San Francisco, they did it in a way that you would expect from one of the world's leading technology hubs. They built Scoot Networks with IoT at its heart.

With more than $5 million in funding and more than 500 bright red scooters in its fleet, Scoot Networks has become the Zipcar of the scooter world. The start-up runs on its software, mobile app, and house-built user interface. Customers create an account and pay via credit card when they reserve a scooter using the app (they can pay by the hour or choose a monthly subscription that lowers the hourly rate). When the system tells customers where to pick up their scooter, they can connect their smartphone to the scooter's dashboard to activate it, find the best route to their destination, and see how much charge remains on the battery.

The app even lets customers reserve parking spaces and extend their reservation, and it can connect them instantly with Scoot support. When they're done with the scooter, they simply leave it at one of many preapproved locations around the city and go on their way. The system uses GPS to inform the company of the scooter's location and that it's available for use by another customer or needs a charge.[1] "Transportation as a service" finds another landing point.

Scoot Network isn't as well-known as Vespa—and that doesn't matter. They're not in the same business. Vespa sells scooters; Scoot sells convenience and urban mobility. Its story reflects the broader reality of IoT and the sharing economy, one of the thriving new economic sectors powered by the confluence of mobile devices, cloud computing, and wireless connectivity. In that context, Scoot Network and the legion of start-ups in similar market spaces are less important for their economic impact than they are as harbingers of a transformation that is already underway—the senescence of the old product-first economy and the evolution of business driven by IoT and inversion.

What Is Inversion?

It is easy to attribute the success of a company such as Scoot Networks to its use of IoT, but its true innovation lies deeper, in the very way it defines its company mission. Scoot's mission is centered around a specific consumer need—easy, convenient urban transportation—and with that mission as a starting point, the company found the most effective combination of products—cute scooters, powerful software, and reliable connectivity—to enable it to meet that need. IoT made it possible to execute the mission, but the true evolution in thinking that Scoot Networks exemplifies is in its conception of its own core competence.

"Over the last five years, we've seen a surge in adoption of transportation innovations, from car sharing to ride sharing and bike sharing. But, a huge gap exists between bikes, buses and trains—which are too slow—and ride sharing and car ownership, which are too expensive and problematic in the city," says Keating. "As demand grows for capital-light, immediate urban transit solutions, we're working with world-class partners to deploy the new transit future: fast, affordable, shared electric vehicles."[2]

We call this radical reconception of company missions *inversion*. As the term suggests, inversion is a reversal of the traditional relationship of product to solution. In the traditional model, a company defines its mission in terms of the products it makes. "Solutions," if considered at all, are afterthoughts: a way to market the products. In a mattress company, traditional product-first thinking would look something like this:

"What's our highest margin mattress?"

"The Amazing Night King Size Deluxe."

"When did we last ship a new version of that?"

"2012, sir. Version 3."

"How can we get our customers to buy it all over again?"

"We could add 20 percent more coil springs and an extra thick top pad."

"What functionality would that add?"

"It would be less lumpy, especially when somebody tries it out in the store."

"That's it?"

"That's it."

"Pick out some new fabric patterns and build me a prototype. I'll get with Marketing and we'll figure out how to sell it."

[Blackout.]

Simplistic, perhaps, but it illustrates the outmoded product-first business model: driven by guesswork, noninteractive, focused on price, and only narrowly innovative. That model still works. Pumping product into a traditional supply chain can and does continue to generate immense wealth and economic activity in some industries. But the rules are changing *quickly*. Customers in the business-to-consumer and business-to-business realms are demanding more of those IoT hallmarks, *choice* and *control*. And while a company such as Scoot Networks may be small, it is beginning to challenge and disrupt the rigid business models of established shipping giants such as UPS and FedEx. Meanwhile, the same dynamic is playing out in a hundred verticals, from health diagnostics to home security to automotive.

Entrepreneurs are unerring finders of hidden value; that's what makes them entrepreneurs. Empowered by IoT, they are uncovering hidden value more ruthlessly than ever. With inversion, they are exploiting it at a speed and scale that are unprecedented.

Inversion is a thought process, a philosophy, and a business model that shifts the conceptual origin point of products, services, and solutions. Rather than thinking about their business in terms of a product line (the "What can we sell to them?" model), businesses practicing inversion start their thought process by focusing on a particular consumer need, and then they apply all available technology—IoT, deep learning, cloud computing, robotics—to reinvent and reimagine how people meet that need. An inverted company is *needs-driven*, not product-driven.

Focusing on needs makes "what if" thinking into a business strategy that frees a company from the constraints of product lines.

Needs-First Thinking

For a company guided by an inverted, needs-first mission, a product development conversation might look more like this:

"What's our customers' biggest concern?"

"Sleeping better."

"Why?"

"Because they don't always get the kind of sleep they need to feel great."

"Why?"

"Because they don't know their optimal sleeping conditions."

"Why?"

"Because they don't have any data. They're guessing."

"Why?"

"Because nobody has given them a tool to get that data."

"Okay, what if . . . we created an array of sensors to detect activity during sleep and then a cloud-based system to analyze those patterns and feed them to our customers on their phones, so they could make changes and feel better?"

"Great idea. We'll get to work."

With more good questions, some research and testing, and an understanding of IoT, this company might drill all the way down to something like Sleep IQ, the sleep tracking system launched by the Sleep Number mattress company. Sleep IQ uses an array of sensors to detect movement during sleep, sends that data to the cloud for analysis, and lets customers see their Sleep IQ score every morning—including metrics such as restful time in bed versus goal time in bed (the amount of sleep the consumer wants to get), average heart rate, average breathing rate, movement during sleep, bed exits, and more—on their smartphones.

How about a needs-first running shoe? You might want the new shoe developed by VIVOBAREFOOT and Sensoria. Instead of building their shoe out of the same old selection of rubbers, foams, and synthetic leathers, this shoe contains a tiny computer and sensors that monitor variables such as speed, pace, cadence, GPS track, foot landing technique, time on the ground, impact score, asymmetry, and toe engagement, allowing the wearer to not just run but analyze his or her gait for efficiency and protection against injury.[3]

The crucial difference between this needs-first approach and product-first thinking is the freedom from the constraints of a product line. The product-first mattress company might also begin its conversation with "sleeping better" as the customer's main concern—but halfway through the conversation, will only accept "build a better mattress" as an answer. The needs-first approach allows the company to take an entirely different path, one that leads it to pivot away from the mattress and toward the need to actually measure the customer's sleep quality. The product-first mattress company is only comfortable building mattresses, but the needs-first sleep company will sell both mattresses and sensor arrays, and both fit comfortably within the mission.

The Characteristics of Inversion

The inversion paradigm for business features these five characteristics, all of which are evident in the mattress example:

- *Inversion begins with satisfying needs or enhancing experiences.* An inverted company sets out to reinvent how its customers meet certain needs or experience part of the world. This removes the limits associated with selling a single product or product line and opens up additional opportunities adjacent to the customer's need.
- *It's based on value delivered, not things sold.* An inverted company defines its core competence in terms of a value brought to the customer rather than a product line: "We help people sleep better" rather than "we make mattresses."

- *It's technologically omnivorous.* An inverted company uses all available technology to pursue its defining customer needs, unconstrained by a product line. When a need is identified that falls within its core competence, any available technology or product may be harnessed to meet that need—including offerings of other companies.
- *It's connected.* An inverted company uses its products themselves to generate immediate and fact-based feedback about how well customer needs are met and what unmet needs remain, not unreliable "product research." And, it uses that same connection to interact in real time with the consumer, becoming an integral and ongoing part of the consumer's experience.
- *It's disruptive.* An inverted company is prepared to do whatever is necessary to fulfill its vision—even if that means pivoting away from yesterday's products or abandoning them altogether. What an inverted company cannot do itself, it gets through partnership or acquisition.

What these characteristics have in common is that they break through the boundaries of traditional products, and that is where IoT comes in. It is obvious how IoT supports the connectedness needed to directly measure customer experience. But IoT also gives businesses unprecedented flexibility to address customer needs by giving engineers and designers new principles for leveraging the inverted paradigm—something we cover in detail over the next four chapters.

When Keating and Riegel started Scoot Networks, they were not looking to create an IoT company per se. They simply had a vision for enabling people to move around San Francisco more easily. But that vision was only realizable because of IoT. That's an important distinction: while the inversion economy is not expressly about IoT, IoT is a key component of its DNA. As we've said, IoT is the essential entry point of inversion into homes and business—sometimes in the form of a smart device, sometimes in the form of a smartphone. It is the inversion value delivery system. There's nothing stopping any new or existing business from reimagining its customers' needs and trying to meet them in a new way; that was the impulse that launched Nest's original smart thermostat. But it's technology—in the case of Nest, IoT—that made that vision both broader and scalable.

A needs-first business looks for ways to reinvent how its customers interact with and experience part of their world, usually in a way that increases convenience, improves choice, saves time, saves money, or all of these. Needs-first businesses create value not by selling to customers but by *empowering* them. Inversion creates businesses that unlock new value in existing markets by adding new layers of utility, creating new markets where none previously existed, and disrupting existing markets and business models with technology.

In one example of that kind of disruption, at the end of 2016, Amazon rocked the world of retail by opening its first Amazon Go location, a storefront with no cashiers and no checkout process. Customers simply walk in, grab what they want, and walk out. In fact, the online retailing giant calls its system Just Walk Out Technology: a combination of computer vision, sensor fusion, and deep learning that automatically detects when customers take products from store shelves, keeps track of the products in a virtual cart, and after customers exit, charges their Amazon accounts and sends them a receipt. It's a startling challenge to conventional retailing because it's not just about moving units, but meeting customers' need for the kind of fast, seamless shopping experience they're used to having on Amazon.com.

The Inversion Advantage

That approach to business confers a powerful competitive advantage: it lowers the traditional barriers that prevent companies doing business in a particular vertical industry from expanding their reach into adjacent verticals. Purely an online retailer, Amazon can leverage its huge edges in product tracking and customer data mining by turning the retail storefront into a physical version of its online store. Value resides in the experience, not the products. In another example, if a home thermostat company such as Nest aspired to extend its product line into an area such as home security in the pre-IoT era, it might find doing so both prohibitively costly and difficult. The company would not only need to design and fabricate new products, but also to create a whole new sales and marketing infrastructure.

But the essence of Nest is that their value is in keeping people in control of their homes even when they are not at home. So while the company had to design a physical product in order to roll out the Nest Cam, the *value*—video surveillance of your home on your smartphone—is a natural extension of its core mission. No new sales and marketing infrastructure was needed—Nest had broken the traditional boundaries between the product categories of thermostat and home security system.

Uber's UberRUSH system is another terrific example of "opportunity adjacency" powered by IoT. The "we move people from point A to point B" company was easily and quickly able to launch a "we move parcels from point A to point B" service when it saw a need that one else was meeting—a transition that few medallion taxi companies ("We operate a fleet of cabs") would ever think about making. By moving quickly, Uber turned it into a business opportunity.

The ability to expand into neighboring businesses easily and cost-effectively to generate new revenue streams and attract new customers is one way that following a needs-first, inversion model gives businesses using IoT a powerful advantage. Inverted companies operate in many ways that are different from product-first companies:

- *Leveraging existing assets to create new value.* In the past, growing a business meant designing and building expensive new factories, growing distribution channels, and operating retail outlets—and then using corporate communications to convince customers that these new lines of business were of greater value than competitors already in existence. The inverted model turns an uphill trudge against market forces into a downhill thrill ride by letting businesses leverage the untapped potential of markets that customers *already understand and perceive as valuable*. The flexibility afforded by IoT lets companies quickly introduce new products to meet the need at the center of the company's mission by developing and deploying new software that changes how its connected devices function. Businesses need not burn large research and development budgets developing and beta testing new products; inversion and IoT empower them to turn the proven, familiar, and even common into competitive assets.

- *Curating, not controlling.* Rather than try to impose their will, needs-first entrepreneurs seek to channel the energy of the marketplace, add value at key intersections, and give users the final control over their experience. This not only frees businesses from the burden of trying to restrict what cannot be restricted, but also wins customer affinity by allowing them to decide which features and benefits they find to be most valuable and worth paying for.
- *Driving solutions with data.* Inversion and IoT combine to collect prodigious amounts of raw data about how customers are actually experiencing a company's offering in situ. A company does not have to guess whether the consumer is satisfied with the experience: it can actually *measure* how much time the user spends pressing different buttons, or it can gather direct feedback as the customer rates the experience. Guesswork and focus groups give way to hard facts.
- *Collaborating.* In the product first world, the relationship between product builder and customer wasn't really much of a relationship at all. There was little or no flow of information back and forth; the product builder had no way to know how its solutions were being used. As we've already pointed out, in such an environment, the creation of a new product featured a substantial amount of guesswork. Inversion, particularly when enabled by IoT, turns solution provider and customer into collaborators working together to continually improve the functionality and utility of the product and the user's experience with it.
- *Exploiting IoT.* IoT and its infrastructure—sensors, actuators, connectivity, microprocessors, cloud computing—are the backbone of inversion. But in this inverted world, even indirectly, they are changing how we perceive and leverage assets that we previously thought of as mere things. Lyft drivers pick up passengers in cars that are, for the most part, not connected to the Internet. However, by virtue of its mobile app and the web of assistive applications and connectivity that power the company's transportation marketplace, Lyft has transformed the simple inert car into a de facto connected device and *dramatically* increased its potential value.
- *Simplifying the stupefyingly complex and dauntingly inconvenient by addressing the total need.* A needs-first company does whatever it takes to

completely satisfy the customer need, looking at all parts of the problem. A product-first company is limited to addressing the part of the customer's process that involves the product itself.

The Sharing Economy

The sharing economy, home to pioneers such as Zipcar and Airbnb and smaller players such as Scoot Networks, SnapGoods, and many, many more, is fundamentally inverted. Generally, sharing economy companies are not pure IoT companies because, while they rely on connectivity to leverage previously hidden value in ecosystems that have always been in public view, they do not rely on the mission-centric connected devices of a traditional IoT player such as Nest. But we contend that they will eventually work pure IoT into their business models, even if they don't already.

However, such companies represent inversion at its purest: starting at the end point with the question, "How can we change how people experience this small sliver of their world and create a profitable marketplace in the process?" It may not even be obvious at the outset exactly which product-first businesses will be disrupted.

This is also the area of inversion with the lowest friction, which is why there are so many players in every sector of the sharing economy, from lodging to travel to cars to shipping to moving. Because there are by definition few (or zero) competitors in a largely invisible market, the primary issue here is finding the seams, the mechanisms of untapped value that are often just below the surface of daily life, and exploiting them. With that achieved, the toolkit for launching a business—mobile app, network of private contractors, strong online brand, third-party resources such as mapping and payment systems—is well established.

Pitfalls

Despite the popularity of the sharing economy in particular, novelty breeds discomfort and slow adoption. So inversion-minded companies in the business of reinventing themselves should expect some resistance from customers facing a thoroughgoing tear-down and rebuilding of a previously familiar experience.

This will depend not only on the audience (young early adopters versus Baby Boomers, for example), but also on the sensitivity of the realm in which IoT engineers are playing. The notion of busy urban intersections where nobody actually stops will probably require more consistent public relations and education efforts to achieve widespread acceptance than would a home irrigation system that can be remotely controlled from a smartphone.

The other danger for the "turn something old into something else" model arises when engineers and designers become so enthusiastic about IoT that they apply it to areas of the economy that don't need it—or where it might actually be unwelcome. "IoT everywhere" may be a hot hashtag, but there will be corners of the economy where it doesn't apply—where data crunching seems reductive and even insulting. A "smart couch" for psychotherapists, giving them the ability to see their patients' heart rates, body temperature, and other signs of arousal in real time during therapy sessions, might seem like a winning idea. But would it really be a benefit in such emotionally charged environments? Such a product could become a distraction for the therapist and feel like a violation to the patient.

Inversion should be about meeting real customer needs, creating valuable experiences, and powering new markets, not about asking, "How can we insert IoT into our existing business?"

Not Just for Start-Ups Anymore

Because they can have a clearer path to scale, sharing economy giants such as Uber also tend to dominate the headlines in the world of disruptive high-tech business. This creates a prejudice among more established companies, especially those in fields perceived as being relatively low-tech, that the connected world, IoT, and inversion are mostly for agile, young start-ups. This could not be further from the truth.

In reality, it's precisely the power and flexibility of connectivity, sensor- and actuator-rich devices, smart data, and cloud computing to transform legacy companies and create trillions in new value that makes inversion possible for businesses of all sizes. Already, "analog" industries are embracing IoT to automate operations, provide data that increase efficiency, upgrade safety, and much more. IoT is already changing such bellwether sectors as health care and heavy industry and is having a powerful economic impact. In fact, according to McKinsey, by 2025, IoT-enabled services and applications are expected to deliver between $200 billion and $1.6 trillion in new value in health care, $900 billion to $1.7 trillion in new value in municipal services such as traffic and power grid management and public safety, and $1.2 trillion to $3.7 trillion in new value in factories and manufacturing in such areas as operations and predictive maintenance.[4]

These are sectors of the economy where start-ups can play, but long-established businesses own the ball and the field. Inversion using IoT can reinvigorate established businesses in ways that differ from those of start-ups. If you run or work for a legacy business that has achieved some measure of success, then you already know how to do what you do with a degree of success. You know how to deliver value, service customers, solve problems, and generate profits. For such businesses, the danger of this era is the unperceived slide into obsolescence—an agile, ambitious, nothing-to-lose start-up blowing past you with a connected, data-enriched solution that gives customers value that you can't match at a speed you can't emulate, with the cool factor of IoT playing the role of salt in the wound.

Established businesses that wish to avoid that slide are embracing inversion not necessarily to create new markets but to take those assets we just described—the things they already do well—and focus them more directly on underlying customer needs. They're using IoT to add real-time data, automation, greater efficiency, reduced cost, or enhanced convenience to what they're already doing that's profitable and popular.

Take venerable tractor maker John Deere. The industry in which it has been a major player since the 1830s, agriculture, has not traditionally been data-rich. Only farmers have had any idea of production, irrigation volumes, and a hundred other variables involved in taking crops from seed to silo, and even then data has been rough and often based on human observation.

Now John Deere is using IoT to layer new value on top of its long-standing business model, leveraging its existing relationships to do so. The company has developed a mobile online platform, JDLink, that gives farmers remote, real-time data on the location, activity, and diagnostic status of each machine involved in their operation, from tractors to combines. Even more ambitiously, the company's operations center uses connected devices to give farmers a wireless stream of information on weather, soil conditions, crop status, irrigation, and a lot more. Given that farming is a precarious business at the best of times, John Deere has made itself invaluable to its customers by helping them reinvent part of how they work—using technology to help reduce uncertainty in an uncertain profession.[5]

Inversion can revive even large businesses that seem, at first glance, not to be IoT-friendly. That's why thousands of forward-thinking corporations in virtually every vertical industry are busy developing their IoT offerings. Some forward-thinking companies are even building businesses on helping other companies make this transition. ServiceMax helps manufacturing companies move away from their product-based business models—making and selling objects to customers—to "manufacturing as an IoT utility."[6]

For example, a solar panel manufacturer working with ServiceMax would cease asking its customers to purchase panels and installation. Instead, customers would subscribe to a "power purchase agreement." Instead of the customer spending tens of thousands of dollars to purchase a solar array—and separate costs to upgrade and maintain it—the customer subscribes to "solar energy as a service." The solar company installs and manages a cloud-connected photovoltaic array that it then monitors 24/7 for power production levels and reliability. The company charges the homeowner a monthly fee that can change based on needs. That is agile, needs-based business that can adapt as the customer and market change.

The Six Business Fundamentals, Revisited

Coupled with the ubiquity of IoT, inversion has the potential to change everything about business, including the fundamental principles we discussed in the last chapter. Let's draw this chapter to a close by revisiting those foundational ideas and seeing how inversion is causing them to undergo their own evolution, and how new fundamentals are emerging as well.

1. Brand

In the old world, brand is primarily about the promise of an experience—the tribal sense of belonging, of owning an iPhone, the anticipation of comfort in flying Emirates, and so on. Now, with open feedback channels and transparency, brand is increasingly equal to the experience itself, a triumph of substance over style. Airbnb's promise of a terrific lodging experience takes a backseat to your actual experience as curated by the company, from the ease of booking to the quality of your lodgings themselves.

2. Competition

Competition is not disappearing, but inversion and IoT are blurring the lines between companies. If New York City–based car-sharing company Car2Go relies on Google Maps to track its inventory of available vehicles, does that mean it's a competitor with Google, a partner, a customer, or all of these? Inversion leads companies to find the technology they need to meet customer needs wherever it is available: if some other company has it, you'd better partner with them or acquire them. In some sectors of the economy, competition may evolve to look more like collaboration; in others, collaborating companies may find it necessary to wall off some IoT operations where they are interdependent, allowing open cooperation in some parts of their businesses while remaining cutthroat competitors in others.

3. Innovation

In the inversion era, innovation can happen at a higher level than ever before. The reason is simple: breaking free of the limitations of the physical world and creating solutions in the digital world means that increasingly, anything really is possible. In that environment, companies and customers alike expect the truly new and groundbreaking; they will not pay for anything less. This makes the act of innovating more challenging, because the bar has been raised, but it also means that engineers, programmers, and designers can really begin conversations by asking, "What if?" and know that there's a good chance they can build an answer.

4. Market Share

As with competition, inversion blurs the lines of demarcation that once indicated market share. When multiple businesses collaborate to provide IoT components and infrastructure to make a complex solution possible—driverless cars on a smart grid, for instance—who owns that market? Does each stakeholder claim a fractional market share? How do we translate that into metrics such as shareholder value and growth? Inversion does more than disrupt industries and business models; it upends some of the basic definitions that define business.

5. Ownership

Again, blurred lines and deep collaboration challenge the concept of ownership, but so do inventories that are completely unconventional. Who owns what in the world of tool-sharing company NeighborGoods? When ownership is unclear, who is responsible for fixing errors, mitigating problems, and addressing legal dilemmas? Even in more straightforward sectors such as utilities, where IoT might be used to deliver data streams from plants and grids, who owns the raw data and the actionable conclusions that cloud-based number crunching provides? The inversion age will require not only fresh business thinking, but also an evolution in concepts of intellectual property and intellectual property law.

6. Value

Of all six fundamental ideas, value comes with the largest degree of change in the inversion world. In the connected future, value will arise from giving the customer the most satisfying experience possible. The trick is that each customer's sense of what is satisfying will potentially be changeable from moment to moment due to the dynamic nature of IoT itself. Versatile, immersive systems that learn about customer needs and act as autonomous agents offer the possibility of constant adaptation and expansion into new services and new points of value.

Value is decoupling from form and features and becoming whatever the customer finds valuable in the moment, and increasingly it's about those C words: choice, control, and convenience.

How to Speak Inversion

In the face of such abundant, disorienting disruption, some things will remain constants. Revenues and profitability will be the chief measures of success. Growth will continue to drive valuation and share price. And product managers, engineers, designers, and entrepreneurs will continue to work together to build the impossible. However, to create value in this new era, they will need to learn a new vocabulary.

Fortunately, we have identified four essential principles—connectedness, composability, recruitability and immersion—that all stakeholders can use to develop and build the next wave of world-changing solutions. Fluency in these concepts will become as pivotal to the inversion age as fluency in metallurgy and mass production were in the age of Henry Ford. The result is a new vocabulary: a new way of talking about and thinking about business and technology in the age of IoT. In the next chapter, we will go into detail about the meaning and usage of this new lexicon—and how fluency in it is critical for building the inverted, connected economy.

3 An Inversion Vocabulary

As we have discussed, inversion is a paradigm for evolving from a product-first orientation to fulfilling needs and finally to creating and curating customer experiences. As a company moves through this process, it ceases to direct all its resources on perfecting products before they ship and begins to direct some of those resources on identifying unmet needs, measuring consumer experience, satisfying new needs through reapplying existing products, and identifying value hidden in the economy in which its services are consumed.

While an inverted mission centers on customer needs and experience, the means to fill those needs and create those experiences still requires physical products. But designing products to carry out an inverted mission requires a new way of thinking about how devices, software, and systems should interact with each other and with human activity. This is not a matter of simply "IoT enabling" existing products—it is about embracing a new vocabulary for product design that connects directly to the needs-based orientation of inversion. In this chapter, we will explain that need.

The Evolution of a Traffic Light

Products come about in part because of how engineers and designers think about their nature. These patterns of thinking shape the final outcome of the work—the product that comes off the production line, the software application loaded onto the computer hard drive. In part, this thought process is driven by assumptions about what the product should do and the role it is expected to fill in the world.

In the product-first business environment, these are some of the more common assumptions about products:

- *They have impermeable boundaries.* A product's purpose and performance does not change, regardless of its environment, the functionality of other products in its proximity, or how it could be used to meet additional needs. A mattress is always just a mattress, even if a need for greater functionality exists and there are other products available to help meet that need.
- *They are inflexible.* Once their designs are approved and their manufacturing specs set down, products are unchangeable. New functionality means a new version, a new box and a new fee. Too often this has even applied to software, whose ones and zeroes are much more easily revised than the molecules of a DVD player or waterproof hiking shell.
- *They do one thing.* Coffee makers brew coffee. Hair dryers dry hair. Restroom towel dispensers dispense towels.
- *They are "black boxes."* Users—and often, the manufacturer—have no way to access data that reveals the product's performance, use patterns, or anything else that could be used to improve the product or the customer's experience.

However, what if engineers and designers began with a different set of assumptions? What if an organization's default mission was not "build more boxes" but developing solutions that meet needs with an eye on eventually creating new experiences for its customers? How would that assumption change the outcome of those engineers' and designers' work?

For one possible answer, we bring you the humble traffic signal. In its product-first form, it's a simple device that is not connected to the Internet. Controlled by local switching elements that read nearby traffic levels and flow through induction loops in the street, it has one function: flash colored lights that instruct drivers to proceed, slow, or stop.

Now, let's approach the traffic signal from a "meeting needs" perspective. At the inverted traffic signal company, engineers don't think of the traffic signal in terms of what it *is* but in terms of the need that it satisfies—in this case, a city's need for safe, efficient traffic flow. From that perspective, this repurposed traffic signal is no longer merely a green/yellow/red light but also a wireless platform.

This platform hosts hardware and software that collect data on traffic patterns, traffic speed, congestion, emergency activity, road conditions, road closures, and more; send that data to the cloud for processing; and continuously give the city information it can use to predict the need for road maintenance, to dispatch emergency services, to raise or lower speed limits depending on traffic density, to switch the directionality of high-occupancy vehicle lanes, and much more. And yes, the signal still flashes green, yellow, and red.

However, introduce self-driving vehicles and the traffic systems developer must create another new thing—one that is no longer oriented on meeting a need but delivering an *experience*. Gone is the green/yellow/red signal; in its place is a sensor-rich beacon that controls the flow of traffic from the four directions that converge on its intersection. Stopping at intersections is now a thing of the past. Instead, the beacon detects the transponder signals of oncoming vehicles, analyzes their speed, determines the adjustments in speed necessary to prevent collisions, and instructs each vehicle's onboard CPU to slow or accelerate to keep traffic flowing in all four directions continuously and safely. Instead of an intersection, we have something more akin to a traffic circle—albeit one that also collects and uploads terabytes of traffic data every second to the cloud for municipal use.

That's more than a new vision for a traffic signal. That is a complete business model inversion. From the product-first perspective, the vendor's business model was "We sell traffic signals." Now, from an inversion perspective, it has become "We sell fast, safe urban transportation experiences." While Google and Tesla develop self-driving cars, societies need to think of intelligent highways—as California did thirty years ago with the California Partners for Advanced Transportation Technology, or PATH,[1] program at University of California, Berkeley that develops ideas and technologies to improve California's surface transportation systems.

A Changing Mission

The evolution from products to needs to experiences is turning a wide range of product categories into connected IoT-powered devices—even humble, utilitarian products such as suitcases. Consider Raden's line of "smart luggage," which the company's CEO, Josh Udashkin, even describes as a "platform." The lightweight rolling carry-on bags not only function as chargers for mobile devices, but their careful composition of mobile app, sensors, and a crowd of GPS networks enables the suitcase's display to provide local traffic and weather information, show security wait times at selected airports, weigh themselves (helping travelers avoid excess weight fees), and even send their owners a proximity alert, something that's especially convenient in crowded baggage claim areas.[2] The suitcase is no longer a carrier for clothes, but a nexus for everything you need to do while traveling.

The microprocessor-powered IoT device on the office shelf, in your pocket, or under the dashboard of a crosstown bus no longer need be limited to a single function, vision, or purpose. It can operate in many contexts simultaneously. In the inversion era, the rules are changing. Boundaries are permeable. Designers can develop objects with multiple intended uses from their conception, and as new use cases present themselves, other designers can find new ways to recruit the functionality of those objects to do new things. Engineers are building connectivity into products and maintaining communication with them after they are sold, seeing how the product is performing and being used, and using that knowledge to develop new services that better meet customer needs. The focus is no longer on things locked into a single state but on *platforms* that *behave*. Instead of asking, "What is it?," we are beginning ask, "What can it *do*?"

However, success in this space is more than just a matter of changing your mission to a needs-based one and then going back to doing product development the way you did before. Products come into being because of assumptions, as we've discussed, but also because of the language we use to talk and think about them. To practice inversion as a business strategy and an engineering model, you need a new vocabulary with which to discuss, design, and deploy the technologies that make the connected world possible.

This new vocabulary is the starting point for building new things, taking things you already have and adapting them to meet new needs, and empowering the customer to use what you give them to satisfy their own needs.

Four New Engineering Principles

As we developed the inversion paradigm and everything we share in this book, we determined that the current engineering vocabulary used in product-first businesses was insufficient for inversion. Designers, engineers, and entrepreneurs who set their sights on creating solutions for the needs-based, experienced-based marketplace must learn to embrace four new engineering principles.

1. *Connectedness* is the ability to connect a device to a data processing platform through the Internet using a communication protocol—usually, in the IoT age, a wireless one. The processing platform can be an onboard computer (such as the microprocessor found in many home appliances or the computers in modern automobiles), a local hub (such as a smart base station used by residential audio or security products), or a remote system (most commonly, the cloud). The ability to connect devices to the Internet wirelessly defines IoT.
2. *Composability* is a quality that enables the customer to combine the functions of multiple connected devices, creating "composite" devices that together perform new functions in ways that meet

The Vocabulary of Inversion

- Connectedness
- Composability
- Recruitability
- Immersion

needs their designers may not have foreseen. Note that the user is not creating new devices any more than an artist creates his subject matter; she is merely assembling them into an overall composition to satisfy her immediate need.

3. *Recruitability* is the capacity of a system to be readily adaptable to additional functions beyond those for which it was originally designed. A recruitable solution typically features an open architecture, externally-facing application programming interfaces (APIs), and function nonspecific controls (LCD touchscreens, voice, cameras, haptics) that can accommodate unanticipated adaptations, including those proposed by third parties, with relative ease.
4. *Immersion* is the principle of composing devices, software, and distributed "fog" or "mesh" computing power to create a seamless constellation of intelligent, environmentally responsive devices with the user at the center. Working in concert, these technologies become a mobile, responsive, immersive personal ecosystem designed to create high-value experiences for the customer—and let the customer control his or her own experiences.

Early innovators of IoT incorporated these principles into their products by fortunate accident, the by-product of science and innovation colliding at high speeds. However, in the future, intelligent devices that succeed in creating extraordinary value for customers and companies will have these properties "baked into" their development from the earliest conceptual stages.

That is the purpose of the inversion vocabulary. By codifying the essential engineering principles necessary to conceptualize and create solutions for the inverted, connected world, we actually create a new lens through which to view business and the world. Existing, outdated frames of reference limit creativity and innovation. The vocabulary we use changes those frames of reference, empowering us not only to express and realize ideas differently to colleagues but to more readily express them to customers as well. That is the purpose of these four inversion-focused engineering principles.

Any business—from disruptors making brave forays into IoT to desperate late adopters struggling to compete—that wishes to remain relevant as this evolutionary process progresses must understand and master this language. Let's examine each part of this vocabulary in greater detail.

Connectedness: Putting the Internet in IoT

Without access to the Internet (or to another device with its own Internet connection), a toothbrush is just a toothbrush. Communication is the general term describing the ability of devices to connect to the Internet or another device through Wi-Fi, cellular data, Bluetooth, or any one of many other communication protocols.

Connectedness makes IoT possible, most often by connecting devices equipped with sensors and actuators to the processing power of the cloud. Adding connectedness enables even ordinary devices that we do not think of as being high tech to become part of IoT. For example, a traditional residential doorbell has two components: the button and the chime. It uses a hard-wired connection to send current from the button to the chime, and that is all it can do. It's a simple machine.

Let's turn the doorbell and chime into IoT devices and connect them to each other via the home's Wi-Fi network. No wiring is necessary; both devices run on battery power, can be easy moved or modified without damaging walls or ceilings, and can be monitored and modified using a simple mobile app. Through connectedness, we've taken an outdated system and made it flexible, convenient, and transparent to both the user and the seller.

How this furthers inversion: Being connected to their products via the Internet allows businesses to receive a continual flow of data about use patterns and device performance that can indicate possible customer needs that are not being met. Connectedness also gives customers greater flexibility in using what they buy, a natural segue into our next principle, composability.

Composability: The Artist

In designing composable devices, engineers give customers greater control over their own experiences. For example, the preceding homeowner would like his doorbell to control other systems: a porch light, a home security camera, and an intercom. Hiring an IoT vendor to do the work, he adds a wireless, Bluetooth-enabled porch light and configures the doorbell to turn on the light. Next, he adds a battery-powered, Wi-Fi-enabled security camera and configures the doorbell to trigger the camera when pressed. Finally, he adds a wireless intercom system that allows him to see and speak to people outside the home using his smartphone. Each device has a virtual counterpart in the cloud; the doorbell, camera, light, and intercom have all become services, each with a virtual software interface accessible via a cloud-based application. The whole system is reliable, easy to maintain, and can be easily upgraded.

Composability serves a "teach them to fish" mentality that turns a company into a collaborator with its customers. Often employing smartphones and mobile apps, customers can continue finding more and more novel ways to combine the functionality of multiple devices—the first stage of the "experience economy" that finds its full expression with immersion.

How this furthers inversion: Composability allows businesses to meet new needs more rapidly. Rather than developing and releasing new products, they can compose new solutions from products that are already on the market—and often, already in the hands of customers.

Recruitability: Take My Functionality . . . Please!

Recruitability means that a device is designed with the capacity (extra sensors and actuators, various connectivity protocols, redundant microprocessors) and/or adaptability (an API that makes it easy to reprogram) that allow its functionality to be *recruited* (or borrowed) to perform functions beyond those for which it was originally designed.

The more ways a device can interact with the physical world, the more ways in which it can be recruited to meet new needs, and the more it can be used to recruit other devices in proximity to it. An example is the smartphone, whose sensors and actuators include a touchscreen, a motion sensor, GPS and cellular location sensor, a camera, a microphone, several pushbuttons, a color display, a loudspeaker, and a haptic vibromotor.

Or consider the Roomba robot vacuum cleaner, a familiar and much-loved product. In its first version, it was just a thing: it could only vacuum a room. But then iRobot thoughtfully released an API so people could recruit the Roomba into other applications. Now, the Roomba became more than a vacuum: it became a mobility platform that could autonomously move stuff around. With this capability, the Public Laboratory for Open Technology and Science in Brooklyn, New York, recruited a Roomba to become a mobile air quality monitor by outfitting it with an autonomous sensor.

Recruitability is central to the inverted, connected world because it moves devices beyond having a single, fixed function to being adaptable to multiple uses in different contexts that may never have been anticipated by their designers. No longer products, devices are *platforms* that can be tweaked or expanded to deliver multiple solutions from multiple sources—including competing companies.

Recruitability is changing the development, design, and engineering conversation from "How do we build a box that does A?" to "How can we make our box as versatile and accessible as possible?" This multiplies the potential benefits that a device can deliver to a customer—and the potential value to the businesses that makes it.

How this furthers inversion: Like composability, recruitability lets businesses capitalize on new opportunities by rapidly deploying new products and services. However, rather than composing the existing functionality of its current offerings, the business can recruit its existing hardware to perform new functions, meet new needs, and create new value—or work with a third-party company to do so.

Immersion: It's Magic

With immersion, we have reached the rarified air where devices disappear—or at least merge into a seamless whole. This is where the vocabulary becomes literal—where connected devices and the cloud services that make them intelligent become so intuitive and ubiquitous that they change how we use our own language.

For example, as little as two years ago, if you were standing alone in your kitchen and you suddenly said, "Alexa, order the pasta maker with the highest customer rating," your family members might conclude you had suffered a delusional episode. But in the context of the IoT and the immersive technology of Echo, what you've said makes perfect sense. At this stage of inversion, we've changed some of the words we use and thus, how we interact with the world around us:

- You tell someone, "I'll WhatsApp my location to you!"
- A gentleman has his car park itself because the only available space is tight and he doesn't want to embarrass himself in front of his date.
- A woman relaxing after a hard day says to her TV remote control, "Please find and play a James Bond film."
- A mom presses the Amazon Dash Cottonelle button next to the toilet paper dispenser in her downstairs bathroom and within an hour urgently needed replenishments arrive at her house via drone.

Immersion is not about creating a single system that delivers all these intuitive functions but a seamless collaboration between systems. Immersion takes businesses beyond the level of delivering products or meeting needs and into the realm of *empowering, curating, and creating experiences* that transform the mundane into something prescient, efficient, and frictionless. That's why we talk about magic. Immersion is what makes the inversion economy seem enchanted.

What might that economy look like? A small business based in a smart office building might outfit its executives with Google Trips so they can keep track of all their travel arrangements via a dedicated Gmail account. When they arrive at a rental car desk or hotel, the app has already messaged them their reservation information, confirmation number, and premium club membership number. Meanwhile, the building is strewn with sensors that detect the levels of supplies that need periodic replenishment—for instance, printer toner—and automatically reorder. The building itself responds to its environment, detecting sunlight angle and intensity and unfurling an array of solar-powered shades to moderate the building's internal temperature and reduce energy costs.

Then there's connected clothing, in which apparel items are data points in the cloud with their own unique digital identities. Global RFID company Avery Denison's Janela Smart Products Platform and IoT data management company EVRYTHING form one of the collaborative teams making this most immersive of technologies a reality, including clothing that messages the wearer personalized discount coupons when she walks into a boutique that sells similar items and locates lost items. Imagine a shirt that senses your body temperature and the muscle contractions associated with shivering and adjusts the temperature in your car automatically or a running shoe that senses a limp in your gait and sends an e-mail to your physician or trainer. That's practical immersion that enhances the quality of life.

How this furthers inversion: Businesses can move beyond delivering individual products to creating high-value, seamless experiences for their customers, allowing them to transition from a business model that focuses on selling units to one focused on delivering choice, control, cost savings, and customer delight.

Security and Privacy: A Critical Part of the Conversation

Because security and privacy are not concepts unique to IoT or inversion, they are not part of the inversion vocabulary. However, the security and privacy risks involved in the IoT and the connected economy are like none that businesses have ever faced. Security and privacy safeguards are the necessary lifeblood of any system design—prerequisites of the system's usefulness and constraints on its scope. Thus, security and privacy are as crucial to successful inverted business as any of these four engineering principles, which is why we are including them in this discussion.

Once you put a device on a network, it is vulnerable to attack. If it's compromised, it can be exploited to mount attacks. One example: the denial-of-service attack against Internet service provider Dyn, which used Internet-connected devices such as printers, security cameras, and baby monitors to take down websites throughout Europe and North America.[3]

Ensuring security for an IoT device is a complex undertaking, and it is difficult to give specific "how to" rules that apply to every device. Each device has its own security considerations depending on what it does, how it is connected, and what is used for. But effective security begins with a disciplined approach to reasoning about security, and for that the following framework (courtesy of the Common Criteria, an international standard for computer security certification[4]) is helpful:

- *Assets*. These are things that are valued by owners and that they wish to protect through adequate security. In the IoT world, this refers not only to valuable data but physical objects that can be sensed by the sensors and acted on by the actuators. For example, if a home security system includes a camera, then everything visible to the camera becomes an asset that requires protection—you don't want hackers spying on your house. For a self-driving car, anything the car might hit becomes an asset requiring protection.
- *Threat agents*. These are motivated, capable adversaries who wish to abuse or damage the assets. Two different threat agents may have different capabilities: a teenage would-be hacker in his basement likely has different capabilities than a dedicated governmental cyberwarfare department does. Two different threat agents may also have different motivations: a terrorist who wants to cause visible damage and take the credit has a very different motivation from a corporate spy who wants to steal secrets from a rival and remain undetected. Motivations are not necessarily malicious. A common security threat is simple user error, where the motivation is benign but where a poorly designed system turns an erroneous action into a threat. In considering security, it is important to identify plausible threat agents and consider both their motivations and capabilities.
- *Threats*. Threats are specific modes of attack that threat agents use to exploit vulnerability in the assets. In the IoT, the communication channel is an obvious target. If a threat agent can gain unauthorized access to the device's API, it is then free to use the sensors and actuators to abuse or damage the assets. The computer is also a target, as the attack on Dyn illustrates. The attackers compromised network-connected cameras and printers, but they didn't take pictures or print anything; they exploited embedded computers to attack other computers. The sensors and

actuators may also be targets of attack, not only because of the possibility of direct use by the threat agent, but also if the sensors and actuators have vulnerabilities that allow the threat agent to modify their behavior.

- *Countermeasures*. These are the steps that designers of systems take to close off the lines of attack and therefore minimize the risk posed to assets by potential threats. Specific countermeasures depend on the circumstances, and often need to be deployed in combination, but very roughly can broken down into four areas: authentication, authorization, confidentiality, and integrity. Authentication means confirming that you are who you say you are, as when a website validates your password when you log in. In the IoT context, authentication would be used so that the device confirms the identity of the system attempting to gain access via its API. Authentication would also go in the opposite direction, so that the external system confirms that it is talking to a legitimate device. Authorization means that once the system knows who you are, it determines what you are allowed to do or see or what actions you can take. In the IoT setting, authorization determines which API actions are allowed to each authenticated user.

Confidentiality refers to using encryption so that unauthorized third parties can't listen in to the communication between a device and the external party. Integrity refers to ensuring that data is not tampered with, typically using a cryptographic message authentication code that is computed from the protected data—if the data is altered, the message authentication code no longer verifies. In addition to these four types of countermeasures that seek to create a secure system, designers also need to be mindful of the possibility that faulty computer code may unintentionally introduce vulnerabilities. There are many well-known software flaws of this kind, and designers need to be aware of them and implement rigorous quality assurance procedures to minimize their occurrence.

A framework such as the Common Criteria helps designers anticipate security problems and prevent them before they occur. However, even the most well-designed system may be found later to contain a flaw. Designers need to anticipate the need to upgrade firmware to close security flaws once they are discovered. One vulnerability with IoT devices is that upgrading firmware is not always easy for users; on other occasions, users fear that an upgrade will fail and leave the device inoperative. Designers should anticipate this and make firmware upgrades easy and reliable. (In chapter 5, we will see how the Intelligence of Things stage makes this easier for intelligent devices.)

Privacy

Privacy is really a special case of security, where the assets that require protection are personally identifying information (PII). Your name, address, phone number, and Social Security or taxpayer ID number are all forms of PII. In the context of IoT, especially for IoT devices in the home, PII can also include any information in the surrounding environment that a person would consider to be private. Any security threat that affects PII also becomes a privacy issue.

What makes IoT challenging is that it breeds new types of PII. A camera in the home is an obvious source of PII. However, other forms of information have the potential to become PII. For example, consider a toll pass mounted on the windshield of your car. The fact that your car passed a particular tollbooth on a particular day is not by itself particularly sensitive. But if such data is accumulated over many months, it may be possible to develop a very detailed picture of the driver's comings and goings, and the data becomes PII.

When working in IoT, consider whether what you are doing has the potential to create information that people or governments will consider PII. If so, think carefully about how you will protect that information, who has access to it, whether any of it will be stored, and how it will be used.

The transformation of things from merely connected devices (the Internet of Things) to collaborating, need-satisfying devices controlled by software (the Intelligence of Things) and finally to invisible components of a seamless experience (the Immersion of Things) is the story of moving from making products to satisfying needs to orchestrating experiences. It is the subject of the next section.

II The Evolution of Things

4 The Internet of Things

Breaking the Boundaries of Form

In chapter 3, we unveiled an "inversion vocabulary": a set of engineering principles comprising four terms—*connectedness*, *composability*, *recruitability*, and *immersion*—that will enable engineers and designers to more effectively think about, conceptualize, and build products that fit the needs-first orientation of inversion. By building these characteristics into the products they make (or by looking for these characteristics in the products they acquire), companies will be positioned to meet needs in the deepest possible way, fully realizing the potential of inversion.

IoT provides the tools to achieve these characteristics: sensors, actuators, computation, and communication. However, merely connecting a device to the Internet does not by itself achieve the characteristics necessary for building products around the objective of meeting customer needs. Many businesspeople, misled by hype, regard IoT through that narrow lens. In doing so, they restrict their ambitions and fail to capitalize on the full potential of IoT to create experiences, recruit customers as collaborators, and unlock hidden value.

The following three chapters will explain how businesses can leverage IoT technology and inverted thinking to design devices and solutions with the four characteristics so critical to inversion—devices and solutions that meet customer needs and enhance customer experiences. Understanding the evolving nature of IoT is critical to successful implementation of the inversion business model. It makes clear how IoT and the technologies that underlie it are making it more possible than ever not only to meet existing customer needs, but also to identify untapped areas of value and capitalize on them quickly, anticipate new customer needs before they become acute, recruit customers as collaborators in developing new solutions, and curate experiences that satisfy customer needs in ways that not long ago were possible only in science fiction.

This is a story told in three stages: the Internet of Things, the Intelligence of Things, and the Immersion of Things.

Sensors, Actuators, and Computation

In section 1, we discussed how humans' first use of found objects (function follows form) led to the craft of design and manufacturing (form follows function), culminating in digital technology as a design method of unprecedented versatility, especially in the form of a microprocessor. IoT picks up this evolution of how humanity has adapted the physical world to meet its needs.

But first, let's look at a modern digital device—say, a modern coffeemaker—and see why it falls short of being part of IoT. The components that make the coffeemaker what it is—an automatic device that brews coffee—can be broken down into three groups:

1. *Sensors*—Devices that measure the physical world and turn those measurements into digital data. In the coffeemaker, the sensors include a water temperature sensor, another temperature sensor in the heating plate, a water level sensor, and the buttons that the user presses to start and stop the brewing process.
2. *Actuators*—Devices that turn digital data into changes to the physical world; the converse of sensors. In the coffeemaker, the actuators include the heating element that heats the water, another heating element that heats the coffeepot, a valve to control the water flow, and the indicator lights on the front panel.
3. *Computation*—Digital logic that connects the sensors and actuators in a particular way to achieve the designer's intended purpose. In the coffeemaker, digital logic responds when the "Start" button sends a digital signal indicating that is being pressed. In response, the logic applies power to the water level sensor and receives a reading indicating whether or not there is sufficient water to brew

coffee. If the sensor indicates that there is sufficient water, digital logic applies power to the water heating element and starts receiving temperature readings from the water temperature sensor. As the water reaches boiling temperature, the logic opens the valve to let the water flow through the coffee, then cycles the heater on and off to keep the water from getting too hot or cold and thus maintaining the ideal brewing temperature until all the water has flowed. At that point, the logic turns on the "ready" light and turns off the water heater. The coffee is ready to help its bleary-eyed owner wake and get ready for the day.

In the early days of the digital coffeemaker (or another similar device), that digital logic was custom-built. Designers had individual logic gates that they had to wire together to do the things they wanted the device to do. When the microprocessor came along, that changed. Now, instead of individual logic gates, you could just put a microprocessor chip in the coffeemaker and write a little bit of software that would tell the microprocessor what to do when the Start button was pressed, when the water reached ideal brewing temperature, and so on.

This represented a quantum leap in the complexity of the systems designers could build. Now, instead of employing the difficult, expensive method of assembling hardware components such as logic gates, you could use software to manage complexity. For example, rather than build a device with just a few buttons, you could design a system with a keypad and digital display that could perform several different functions, such as the microwave oven that displays the time of day when the display isn't used to show the cooking time.

However, while the microprocessor offers a simpler, more cost-effective way to manage existing functions and design new ones, the result is still an inflexible system. Functionality is fixed, because the software that runs this coffeemaker performs a narrow set of predetermined tasks and can't be easily modified or upgraded. In fact, engineers call it "firmware" rather than software, because while it may seem "soft" while the engineer is designing it, before the product is manufactured, the code is burned into a read-only memory and can't be changed. Even devices that have upgradable firmware still call it firmware, because from the user's perspective, the functions don't change. It may as well be hardware.

At this point, we still have a coffeemaker that is no different from the one many of us have in our offices. This is not IoT yet. But it's important to understand that this digital device still features three of the essential components of IoT: sensors, actuators, and computation. But IoT adds a fourth ingredient to the mix: *connectedness*.

Getting on the Network

Microprocessors have been part of common household and office devices since the late 1970s. The Internet has been available to the public for more than two decades. But in the mid-1990s, most people didn't have Internet in their homes. They had dial-up modems running at painfully slow speeds of perhaps 56 kilobytes per second. Their computers could dial in to a modem pool at an Internet service provider and get on the Internet, but the home itself was never on the Internet.

Then broadband was widely introduced around 2004.[1] Now, through a digital subscriber line or Ethernet cable, people actually did have an Internet connection—a legitimate IP address—in their homes. But because it was only available through wires, connecting devices with microprocessors to the Internet wasn't practical. However, two developments made connecting computers to the Internet practical and desirable. First, people began to get Wi-Fi in their homes. Second, Internet providers began charging customers a fixed price instead of charging them based on the amount of data that they use. Adding another computer to the home Wi-Fi network did not incur any additional cost. In engineering terms, the marginal cost became zero.

Ten years ago, if you had said, "Let's put the coffeemaker on the Internet because it's got a microprocessor inside," people would have said, "Okay, but why? Who's going to want it? No one wants to plug an Ethernet cable into their coffeemaker, they don't want to pay extra for the bandwidth, and it will double the cost of the coffeemaker." Now all those barriers are gone. Now the question is, "*What if* we connected the coffeemaker to the Internet?"

It is now possible to connect everyday devices to the Internet because, in addition to sensors and actuators, they have computers inside them. Most people, unless they're engineers, don't think of ordinary devices as having computers inside them because their firmware makes them indistinguishable from devices without computers. But when you add the microprocessor, it's not a big leap to connect the device to the Internet as long as it contains the necessary sensors and actuators to turn software instructions into action. And with the mobility of cellular data, engineers can not only connect stationary devices such as coffeemakers, but also mobile devices. This has allowed designers to add a wide range of personal items to IoT.

There is a wireless communication technology ideally suited to virtually any IoT application and physical environment. Wi-Fi and cellular data are the primary means for long range wireless connectivity. Protocols such as Bluetooth are intended for a radius of a few feet, or what's called a personal area network (PAN). But for a local area network or wide area network, Wi-Fi and cellular data are the solutions of choice. Generally, cellular data is more expensive, but it has the advantage that it is everywhere and ideal for users who are in motion, while Wi-Fi requires an access point and is better for users working in a closed environment such as a home or office.

Connectedness

Communication provides the piece of the new inversion vocabulary that makes devices with sensors, actuators, and computing part of the IoT: *connectedness*. The implications of this single step are profound and can be built up in stages. As we saw in chapter 3, connectivity is one of the characteristics businesses need to practice inversion, gaining the ability to maintain a relationship to their products after they are in the customer's hands.

There are early examples of this that even precede the Internet as we know it today. The very first satellites launched into orbit in the space race of the 1950s and 1960s had radio links so that mission control could monitor them remotely (telemetry) and make adjustments as needed. Those satellites had all four ingredients of an IoT device: sensors, actuators (such as the retro rockets), computation (very crude compared to today's microcomputers), and communication, the radio link to Mission Control.

Back then, there weren't many things that had all four of those ingredients. But connectivity between a business and its product started to expand outside of government into the commercial world. Still, it was limited mostly to expensive, mission-critical devices, such as aircraft engines or MRI machines. Often these required specialized, dedicated networks. For example, 1978 saw the launch of a specialized communications network called the Aircraft Communications Addressing and Reporting System (ACARS), which uses short messages sent via radio or satellite to monitor the mechanical status of critical aircraft systems while an aircraft is in flight.

That is more than a tool for maintenance and safety. When Malaysia Airlines Flight 370 disappeared on March 8, 2014, investigators were able to estimate where it went down because the jet was periodically reporting its engine status through the ACARS network. By noting its position when it made its last ACARS transmission, officials were able to provide search crews with coordinates to guide their search for the plane. However, in the wake of the tragedy, some critics have asked why aircraft need a physical black box at all. A "glass box" (an idea described in 2010 by Krishna Kavi)[2] with data stored in the cloud, would enable instantaneous access in the aftermath of an incident, potentially saving lives in the process.

Rolls Royce's jet engine division employs its Engine Health Management system in much the same manner: to track the condition of and predict failures in engines operating around the world using onboard sensors and live satellite transmission. With Engine Health Management, the company outfits its Trent engines with up to twenty-five sensors that detect vibration, temperature, speed, pressure, and flows. The system transmits the sensor data to the ground using the ACARS network, and then a global ground network sends the data to the destination, where specialists can analyze it.[3]

So the idea of connecting devices to the businesses that make them is not new. But until fairly recently, it required specialized equipment. But with the public Internet infiltrating every business and phone and the ubiquity of inexpensive wireless communication through Wi-Fi and cellular data, it is now possible to extend connectivity to everyday devices. The goal, however, remains the same: keep businesses connected to the device to make sure it is functioning properly and delivering value to the customer.

A good example is the home continuous positive airway pressure machine that helps sufferers of sleep apnea sleep better and avoid momentary cessations in their breathing by increasing air pressure in the throat. Today, those machines are fitted with sensors and cellular data link that allow third parties to monitor their use. The machine measures how much air pressure is used and how many times the patient stopped breathing momentarily. Devices use cellular or Wi-Fi networks to send that data to a central monitoring hub every night, allowing the user's doctor to see how the treatment is progressing.

Insurance companies also use that data to ensure that patients are complying with their sleep apnea treatment. If the data show that the patient is not using the continuous positive airway pressure device at least twenty-seven nights out of every thirty, the insurer might call them and say, "If you don't start using the machine, we will stop reimbursing you for it." However, while this stage gave businesses and institutions a great deal of useful, actionable data that was often beneficial to the consumer, little if any of it was visible to the consumer, nor did it give the consumer any additional control over the devices they were using.

The Smart Water Grid?

Cities are becoming smarter. Traffic, parking, and utilities such as water are due for major improvements in efficiency and service. Water is an important resource that is increasingly under pressure due to growing populations, increased industrial use, aquifer depletion, deforestation, and wastage. According to the European Environment Agency, water leakage in cities can account for 5 to 50 percent of total output.[4]

Is it possible to create a "smart water grid"? Recently acquired by Xylem, Singapore-based Visenti is already doing so. Founded by three MIT postdoctoral researchers—Ami Preis, Mudasser Iqbal, and Michael

The Smart Water Grid? (continued)

Allen—the company relies on the method seismologists use to find the epicenter of an earthquake: monitoring high-pressure transients. Water works the same way: when water is pressurized and there's a leak, a transient goes out, like a wave. Visenti uses that transient to triangulate the location of a leak in a water system.

"Visenti puts sensors in different parts of a municipal water system, and then we precisely track the pressure transients, taking more than 250 readings per second," says Ami Preis, cofounder and comanaging director of Engineering Systems. "We use the cloud to correlate the locations of these transients—it's very hard to figure out where a pipe is leaking, because it's a linear system—and determine the location of the valve that needs to be shut off in order to isolate the leaking pipe. It's real-time pipe breakage detection."

In India, water dictates the fates of millions of farmers. Failed crops due to a bad monsoon can lead to farmer suicides. Can water be managed as a resource? Vassar Labs says yes. The company works with India's state governments to monitor water consumption using their own sensors and government sensors. This is pure IoT. The company does remote monitoring and uses real-time dashboards to release water from a dam, control a canal, display the level of water in a tanker, or encourage machine-to-machine (M2M) sharing of water based on need. Sensors include canal level measurements with on-contact radar or ultrasound (depending on depth). They measure flow rates in canals using Doppler radar and have level sensors before and after branch points like tee junctions. In closed pipes, measurements are done with ultrasonic clamp-on flow measure devices. The data is uploaded over a cellular network, and if the cell network isn't available, text messaging is used.

"Analysis is done in the cloud to infer water flow," says Prasad Putta, founder and CEO. "This turns out to be extremely complex. You have to figure out whether a crop has enough water. You do that by comparing field capacity (excess water held by the soil after water is drained away), and the wilting point (the moisture level at which the plant wilts). This changes from plant to plant. We measure transpiration (how much water escapes through the plant), and using all that, we can predict how well or poorly the crop is doing, where plants are stressed, and where water needs to be delivered."

Connecting the Consumer

The next stage of connectedness is to give the ability to monitor and control devices not only to business, but also to the consumer. A very early example of this is the telephone answering machine (if you're too young to remember the time before voice mail, please bear with us). It was a recording device that happened to be connected to a communications network. Even in an era where most people didn't have communications networks in their homes, everyone had a home phone line. You didn't have to run a new cable or purchase any kind of new network. You would call your phone number while away from home and pick up your messages by entering an access code. The device would play back the cassette, so it had an actuator as well.

Today, with Wi-Fi and cellular data being ubiquitous and cheap, you can think about any device in the home being connected even if communication is not its primary task—an important cognitive leap. For example, you can control your home alarm system remotely, so that if you forget to turn on your alarm before you leave home, you can switch it on remotely from an Internet browser or an app on your smart phone. Legacy security companies such as ADT missed this opportunity, and start-ups such as SimpliSafe have swooped in to take advantage. SimpliSafe offers customers a wireless, multicomponent solution that goes far beyond basic monitoring for home intrusion. The company's top-line package includes devices that can detect plumbing leaks and impending pipe freezes as well as smoke and carbon monoxide detectors, enabling the company to address needs beyond the purview of traditional home security. SimpliSafe has inverted its business model from "home security" to "selling safety and peace of mind."[5]

The Nest Learning Thermostat is a large-scale example of the potential of remote monitoring and control. When you connect the device to your home Wi-Fi network, you can access it remotely to turn up the heat or air conditioning while you're driving home from a vacation. In winter, you can check that your heat is still working while you're at work, so your pipes don't freeze. This is a good example of a company connecting a device with sensors, actuators, and microprocessors to the

Internet and using that as a means to steer its business model toward inversion.

While giving its customers the ability to remotely monitor and control their home's temperature, Nest also leverages the benefits of communicating directly with its products. The company can monitor its thermostats remotely to determine whether they are cycling the way the customer expects them to. If for some reason, a device tells them that the temperature in a home is sixty degrees while it's set at sixty-eight, the Nest thermostat can automatically adjust the temperature or let the consumer know that his heating system might be faulty. The company can confirm that the thermostat is functioning correctly, learn the consumer's preferences for his home environment, and help the consumer in ways a traditional thermostat manufacturer cannot.

Tesla also supports remote software updates and maintenance. That capability may have prevented a potentially disastrous recall of Tesla Model S vehicles featuring the company's Autopilot driver assist system after a fatal 2016 crash in which the driver, assuming that the system was fully autonomous, took his hands off the wheel. Tesla was able to satisfy investigators from the National Highway Safety Administration by issuing a software update that activates an alarm if drivers remove their hands from the steering wheel during Autopilot driving. If the driver fails to take the wheel within 15 seconds, the vehicle shuts down.[6]

As we'll see in chapter 5, Nest and Tesla have both gone far beyond this sort of simple connectedness.

Connecting Businesses

Consumers are not the only beneficiaries of connectedness. IoT allows businesses, hospitals, universities, and other entities to benefit from the ability of connected devices to gather and enable the instantaneous analysis of data. One example of this is resource monitoring, in which IoT is deployed to continuously track the power and water use and climate of a building, campus, or municipality. By enabling 24/7 monitoring of consumption and anomalies, businesses conserve resources, increase efficiency, avoid downtime, and reduce costs.

The creation of "smart buildings" is one of the most benefit-rich examples of IoT currently enjoying a high rate of adoption. Buildings are complex ecosystems with heating, ventilation, and air conditioning systems consisting of chillers, boilers, ducts, vents, and air handlers. They also feature elevators, parking lots, electric systems, water systems, security systems, safety equipment, and more. Unfortunately, building energy systems are not terribly efficient. In the United States, buildings produce about 38 percent of total carbon dioxide emissions and consume more than 70 percent of all energy generated. Very often in buildings, heating and cooling are on at the same time, which is grossly inefficient. To make matters worse, sooner or later a component breaks, but the heating, ventilation, and air conditioning system doesn't know it's broken. It continues to operate at suboptimal capacity and efficiency, which wastes resources and raises the cost of heating and cooling.

The irony is that most of the vents, fans, vanes, and pumps have controls equipped with sensors, but the building operators don't use the sensors. KGS Buildings, a company founded out of MIT, uses its Clockworks system to collect the data from these sensors and use that data to generate an internal view of the system including warnings, faults, and health checks. Cofounder Nicholas Gayeski says, "We collect that data from over one thousand buildings, including more than ten thousand boilers, chillers, air handlers and pumps for hospitals and universities." Clockworks does everything in the cloud, so the power to process the data is virtually unlimited. The implications for more sustainable buildings are significant: greater efficiency and cost savings, lower energy use (which is important for sustainability, especially for buildings that run on alternative energy sources), and more consistent comfort for users.

This application of IoT isn't limited to commercial buildings, either. Homes can also become "smart." A typical home has a lot of air vents but only one thermostat, so it has only one heating zone. You may not want to cool or heat the entire home; you may only want to heat the upstairs. But you can't. The question asked by Ecovent was, is it possible to create local heating or cooling zones in a home with only one thermostat? That is what we call "fine-grained control."

The Ecovent system uses a retrofitted smart vent that sits on top of heating vents, allowing it to turn airflow on or off, creating microzones like the personalized climate control in an automobile. The system also places pressure and temperature sensors in wall units, along with actuators that can move the louvers in individual vents and registers to further adjust heating and cooling levels. The company is now getting about forty million sensor readings a day, and its goal is to solve the comfort equation—creating micro heating and cooling zones everywhere in the home.

This is what we call "insertology": insert IoT into an archaic system and make it quasi-connected. Most buildings were designed before instrumentation became inexpensive. But while the heating, ventilation, and air conditioning operations for many buildings are still stuck in a model established forty to fifty years ago, you can introduce IoT and wireless communication and create a solution that puts some intelligence in these environments.

However, while these examples present important practical benefits of IoT, it's vital to retain some healthy skepticism about the real capabilities of both industrial and consumer IoT, advises Dave McLauchlan, CEO and founder of IoT resource monitoring company Buddy, based in Seattle and Adelaide, Australia. As enthusiastic about the applications of IoT as McLauchlan is, he's also cautious about the overwhelming hype surrounding it—a good message for companies considering entering the connected world who run the risk of focusing on style, not substance.

From where I'm sitting, IoT is massively overhyped. I think that's had a number of outcomes already. The biggest one is that everyone is trying to get into the space. Everyone is trying to apply IoT to their business. What isn't happening, though, is ROI. A lot of folks are getting into the space because it seems like the hot thing to do, and they're doing so without looking first at the problem they're trying to solve. I think 2017 and 2018 will be the years where we pivot from introducing IoT-related technologies into the enterprise because hype says we should, and pivot to a place where we are employing IoT technologies because they offer the best solution to a problem.

Composability: Connecting Devices to Devices

The leading IoT players are leveraging the key functions of connectedness—remote monitoring and control—to deliver value today to both businesses and consumers. Remote monitoring and control by the business serves the inversion paradigm because it allows the company to know whether it's meeting needs in the marketplace—and in some cases, to better meet those needs. However, providing connectivity to the user also serves the inversion paradigm because it gives users new ways to meet their own needs.

If a consumer wants to make sure his home is safe at all times from intrusion, providing a panel near his garage door that he can use to switch on his security system helps him satisfy that need. But if he can also switch on the alarm from his smartphone when he forgets to use the panel, his need is satisfied more completely than if that capability were not available. So connectivity fits into the inversion paradigm in two different ways. It serves the inversion model by allowing businesses to know how well their products are meeting needs instead of guessing. Providing connectivity to the user himself expands the way needs can be met by empowering the consumer to invent new ways to satisfy his own needs with the product.

At this point, we are beginning to explore the second characteristic in the new inversion vocabulary: *composability*. We are connecting devices not just to computers, but to each other, and in doing so, giving them the ability to do new things.

Consider our coffeemaker once again. By itself, it makes coffee. Add communication, and it is now possible to control the coffeemaker remotely. But the power of this is not in a human's ability to control the coffeemaker from a distance; it's in the coffeemaker to be controlled by another device. For example, say you want your coffee to be ready twenty minutes after you wake up. If you wake up every day at the same time, you can set the timer on your coffeemaker to twenty minutes after the time you set on your alarm clock. But what if you wake up at different times each day? What you really want is to set the time on your alarm clock and have the alarm clock tell the coffeemaker when to make the coffee. Or for your smartphone or wearable device to tell the coffee machine when to start brewing.

Connecting these two devices results in something that is more valuable than either one alone. Instead of having an alarm clock and a coffeemaker, you have a "get you going in the morning system," that includes getting you up at the right time and making sure the coffee is ready to drink the moment you stumble down the stairs. If you wanted to get really ambitious, you could also compose your array of connected devices to have "Reveille" playing on the stereo when you get downstairs, have the microwave turn on to cook the oatmeal you put in there the night before, and have your car start twenty minutes after the coffee is made so it's warmed up and heated by the time you get in.

Composability occurs when devices communicate with each other, allowing their users to mix their functionalities into different combinations that meet needs in new and different ways. This is a vital characteristic of the inversion model because it opens up a new way that businesses can help their customers meet their needs. In the past, a company would figure out internally how to compose sensors, actuators, computing, and communication into a new solution and quickly get something to market to satisfy a need. Now, composability makes it possible for the business to deliver a new solution quickly by combining existing components and for *customers* to compose things to satisfy their needs without the business having to intervene.

Take one of the most mundane yet maddeningly retrograde objects in our homes: the light switch. Traditionally, that switch is physically connected to the lightbulb by a pair of wires running through the walls. The switch physically breaks or completes a circuit to turn the lights on or off. It works, but changing the location of the light or switch is difficult, time-consuming, and costly. The homeowners have to tear holes in the walls, turn off the power, disconnect and reconnect wires, remove and remount fixtures, and hope not to electrocute themselves. If you were running a lighting design company, you might conclude that the smartest business model was to keep your customers dependent on your ability to quickly and safely relocate or install hard-wired lights and switches.

But if you were operating on the inversion paradigm, you might decide that you could provide more value and earn more customer loyalty by letting your customers create their own lighting solution. You make the switch and light independent devices connected by Wi-Fi, allowing the customer to change a room's lighting configuration by simply relocating two removable brackets or even a double-sided tape. You add a light level sensor to the system, allowing the user to use a mobile app to set the light to turn on when it gets dark outside. In the same app, you give your customer the ability to set different lights to turn on or off at different times or change the color of that LED. If you're ambitious, you might even include another sensor that reads how many hours a lightbulb has been lit and alerts you when it's approaching the end of its life. This is the sort of thing the Philips Hue enables with Philips LED lights.

The customer now has the ability to create a solution in a near-infinite number of combinations but will still perceive that it was your product and business that made it possible. They will still perceive a company as meeting their need, giving them control, and creating unique value. That fits well with the inversion model.

The Architecture of Connectivity and Composability

With connectivity and composability, communication augments the sensors, actuators, and computation of a digital device. To fully realize the benefits, there are three architectural principles that designers should bear in mind: mediation, the network, and the API. We now discuss each detail.

1. Mediation

The most profound benefit of connectivity is how it leads to composability: the ability to create new functions and satisfy new needs by letting devices communicate with each other. However, the best architecture for this is not to have each device communicate directly with other devices, but rather to have an intelligent *mediator* in the middle.

Look again at the coffeemaker. We saw how composability can be exploited to create a "get you going in the morning" system, with the coffeemaker, alarm, stereo, microwave, and car all working together to start your day. How should those interconnections be made?

The first thought might be to have the alarm communicate directly to the coffeemaker. To do that, you would go to the setup screen on your alarm clock and enter the network address of the coffeemaker. If you also want it to turn on the stereo, you add its network address to the alarm clock settings. What if you want to be able to turn off the stereo once you're downstairs, coffee in hand? Perhaps you program the coffeemaker to talk to the stereo. Such connections are called M2M because each device talks directly to the other device. The problem is, those connections are now scattered all over the house. If you're wondering why the stereo is behaving oddly and want to find all of the devices controlling it, you might have to look at the setup menu on every device in the house.

A *mediated* approach is superior. Each device is configured to talk to the cloud, and an application running in the cloud forwards messages between the devices. This architecture is sometimes called a "star topology" because a picture of the network looks like a star with the mediator at the center and the devices on the spokes. This architecture may seem counterintuitive, but there are many familiar examples. If two people are in the same room texting each other, the text doesn't go directly from one phone to the other. The data packets go from the sender's phone to Verizon's cell tower, then to a short message service hub, back to another cell tower, and then to the recipient's phone. In this example, Verizon is the mediator between the two phones.

In the future, devices such as the Amazon Echo or Microsoft's Xbox might become voice-activated mediators between other devices, allowing users to orchestrate the activity of many areas of the home or office through a single intuitive interface.

Benefits of Mediation

There are many benefits to this approach. For one, the setup complexity is reduced. Each device need only be configured to talk to the cloud. Once that is done, the interconnection between devices is done by configuring the mediator application that runs in the cloud—no need to touch the devices. You can change the configuration from anywhere.

Second, all of the interconnection configuration is in one place. If you want to see all the devices that might be turning on or off the stereo, you can see them at a glance. You can save the complete configuration and restore it with a single click. An M2M architecture, in contrast, is a *brittle* solution, because each device has to be configured with the network address of all the devices it's connected to. If your alarm clock is connected directly to your coffee maker and Sonos home audio system, and you change the settings on any of those devices, you have to find the configurations for each device and change them all in order to regain the functionality you want.

Third, this device-Internet-device architecture lets you have a different type of network connection to each device. The thermostat on the wall might be connected with Wi-Fi, but the smartphone in your pocket uses cellular data, and some other device might use Bluetooth. If the thermostat doesn't speak Bluetooth, it can still talk to the Bluetooth device because the communication will be routed through the mediator. The choice of communication channel is made on the basis of what makes most sense for each device.

Finally, M2M is a security risk. Because each device has a direct connection to every other device, there are too many points of vulnerability, and no easy way to see them all and control them.

A mediated architecture can even exist without the cloud. One of the first products featuring a mediated architecture, the electronic throttle control or "drive-by-wire" system introduced by BMW into its 7 Series cars back in 1988, was an example of this. With drive-by-wire, the mechanical linkage between the accelerator pedal and the throttle butterfly is replaced by an array of sensors and actuators that measure the position of the gas pedal and control fuel flow and engine performance. The connection between the gas pedal and the engine is mediated by the vehicle's onboard computer.

Drive-by-wire lets the vehicle's onboard computer optimize engine performance, fuel flow, air flow, and fuel economy while reducing the number of moving parts that can fail or wear out and capturing data that either the car's owner or maintenance personnel can use to identify, diagnose, and repair problems.

Mediation lets designers balance the benefits of composability against real-world constraints on cost and performance. Mediation in the cloud allows devices to be composed wherever they are, with the configuration accessible from anywhere. A local controller can act as the mediator when composition occurs locally and there are concerns about reliability or response time. (It goes without saying that it would *not* be a good idea for the BMW to connect the gas pedal and the engine via the cloud.)

Even within a single device, a mediated architecture may be at work connecting the various sensors and actuators to the computing device. This becomes important when we discuss recruitability in chapter 5.

2. Choosing the Right Network

Another benefit of mediated architecture is that it frees designers to choose the right communications network for the task. There is often no single answer to the question, "Which network?" For example, an IoT device might connect to a mediation device via Bluetooth, while the mediation device connects to the cloud via WAN using Wi-Fi. However, it is essential to know that there are numerous connectivity technologies, each with its own ideal use. Earlier, we briefly discussed Wi-Fi, cellular, and Bluetooth, but we go into greater depth and detail here:

• *Cellular data (3G, 4G, 5G)*—Mobile phones connect to the global Internet via cellular data, either 3G, 4G, or (in the near future) 5G, depending on the capabilities of the network. That same capability is also available for embedding into other electronic devices. Cellular data is a WAN technology, meaning a device can connect from anywhere in the world where there is a cell tower. This method of networking is especially suited to mobile devices. The cellular network is specifically designed to provide a seamless "handoff" when a device moves from the service area of one tower to the next, so the user never needs to do any reconfiguration. Cellular data is also ideal for outdoor environments where other forms of wireless networking, such as Wi-Fi, are unlikely to be available. The tradeoff is that cellular data tends to be more expensive in terms of data costs, hardware costs, and the energy required to power the cellular radio.

- *Ethernet and Wi-Fi*—Indoors, the most common form of wireless connectivity to the Internet is Ethernet, called a local area network because it operates over a defined area such as a home or office. Ethernet exists in both a wired form and a wireless form known colloquially as Wi-Fi. Wired Ethernet supports the greatest data rates and has the greatest reliability, but it has a tradeoff: a device must be connected to it by a cable. Wi-Fi is wireless and appropriate for a portable device that moves infrequently, but not for a mobile device that moves over great distances while communication is in progress.
- *Bluetooth*—Bluetooth connects one device directly to a nearby device. Most commonly, Bluetooth is used to connect an intelligent device such as a mobile phone to other non-intelligent devices in close proximity—a PAN. Bluetooth is ideal for connections that only need to operate over a short distance—and should not continue to operate over a long distance. For example, if your phone connects to your car's audio system via Bluetooth, you don't want the connection to persist if you walk away from your car. If it did, anyone still in the car could eavesdrop on your conversation. Bluetooth still fits within the DID architecture because one of the devices to which is connects is usually an intelligent device such as a mobile phone.
- *ZigBee*—Bluetooth is designed for close-proximity connections that persist over a long time; ZigBee is designed for transient close-proximity connections. For example, you might have a temperature sensor mounted inside a refrigerated truck that records the temperature over time while the truck is on the road. When the truck reaches its destination, a supervisor waves a handheld device near the temperature sensor to transfer the readings from the sensor to the handheld device and from there to the cloud. The connection between the temperature sensor and the handheld device only lasts for a few seconds. ZigBee is ideal for this application.

There are additional wireless and connectivity protocols that go far beyond the capacity of these four. We discuss those for our engineer readers in the Tech Alert! box.

Tech Alert!

Additional Connectivity Frameworks

We have discussed Wi-Fi, cellular data, Bluetooth, and ZigBee. However, there are many more wireless communication protocols that are not as well-known but just as important to designers and engineers. The first one, IEEE 802.15.4, is an Institute of Electrical and Electronics Engineers' standard, which underlies ZigBee.

A common hope is that one standard will win and that the world will unite around it. We disagree. There are three different dimensions when it comes to IoT devices: range, bandwidth, and power consumption. Each of these wireless standards occupies a different position in that space. For example, Wi-Fi is mid-range and high bandwidth, but relatively higher power, best suited to devices such as IP cameras. Low power WAN (LPWAN) is very high range but very low bandwidth and low power. It is suited to devices that can last a long time on battery power. Each of these standards occupies an important niche. Instead of trying to unify them, the cloud architecture embraces them all, tunneling security through them to an avatar.

The following is a quick overview.

IEEE 802.15.4

This is a low-level wireless PAN standard managed by IEEE for communicating between IoT devices. Underlies ZigBee and 6LowPAN. It offers lower data rates and lower power than 802.11 (which underlies Wi-Fi); it is better for multi-device communication than 802.15.1 (which underlies Bluetooth).

6LowPAN

This is Internet protocol version 6 (IPv6) over low-power PANs. As the name suggests, it is designed for IP traffic over lightweight wireless networks, including "mesh" networks. Works on top of 802.15.4 as well as other communication protocols.

LPWAN

LPWANs are a collection of narrowband cellular technologies that achieve wide-area reach with IoT devices, in the range of kilometers. The focus is on very small packets and very low power, allowing devices to communicate with gateways or base stations at long distances. Examples include the LoRa Alliance, Ingenu, and SigFox. The 3rd Generation Partnership Project, the standards body for Long-Term Evolution, is working on a couple of cellular-based solutions but has not yet converged.

ISO 18000 6C

The Generation 2 Electronic Product Code, or EPC Gen2, standard for RFID (also a GS1 and International Standards Organization standard) is a protocol for extremely low-power, low-range sensors that can measure parameters such as temperature in addition to reporting ID. Although not conceived as a transport protocol, it can play the role of one for sensors and actuators—such as shelf displays in retail.

There are also several protocols and frameworks for passing information back and forth regardless of whether the connectivity is wireless or wired. A few important approaches include:

CoAP

The constrained application protocol is an Internet Engineering Task Force standard designed application layer for machine-to-machine communications. Similar to the hypertext transfer protocol (HTTP) but lighter weight for embedded devices, CoAP uses a user datagram protocol (UDP)-like transport protocol and a simple command-set (not unlike HTTP) and uses representational state transfer communications. CoAP runs mainly on IPv6. IPv4 is possible, but it is not recommended. Devices in CoAP can communicate directly instead of going through a broker, as with MQTT (formerly Message Queue Telemetry Transport).

MQTT

Deriving its name from message queues, MQTT is also a machine-to-machine communication protocol, but one that primarily uses TCP. UDP is possible but not ideal. Messages go through a broker in a publish/subscribe mode. MQTT uses IPv4 and IPv6. TCP is heavier than the confirmable UDP approach of CoAP, but MQTT has a strong user base.

Web of Things

Web of Things is similar to CoAP but based entirely on World Wide Web ideas: representational state transfer and HTTP, WebSockets, Java Script Object Notation, and the Semantic Web. CoAP evolved for very constrained devices. As the computing power in devices increases, the compromises of CoAP (such as UDP and others) may not be necessary. Web of Things standards, frameworks, and tools are mature and well-understood.

Thread Group

A consortium building on 6LowPAN and backed by industry heavyweights such as Google and ARM, this alliance is gaining momentum. Initially competitive with ZigBee, an alliance seems to be forming between the groups. Since both Thread/6LowPAN and ZigBee use 802.15.4, this is

Tech Alert! (continued)

a hopeful trend. Thread also works across Wi-Fi and Bluetooth because of its reliance on IPv6.

Open Connectivity Foundation/IoTivity
Formerly called Open Interconnect, or OIC, the Open Connectivity Foundation is another alliance but with different architecture. Based on CoAP, the IoTivity "full stack" framework also has a plug-in for MQTT. Members include Dell, Intel, Samsung, and Atmel.

Paper Towels as a Service?
To the non-engineer reader, that may seem like Greek. However, each network protocol has potential uses in the context of IoT and inversion. For example, LPWAN is powerful technology for linking together thousands of sensors. Consider the automatic paper towel dispensers commonly found in the restrooms of commercial buildings. Today, restocking is done by the custodial staff, who must check each dispenser's towel level. It is inefficient and time-consuming, and paper is wasted. Instead, what if these dispensers had simple paper level sensors? What if all the dispensers in a building or campus communicated through LPWAN to report their contents to a single server?

The replenishment of paper towels would be "on-demand" and efficient. Custodial staff would only visit the restrooms that needed paper. Furthermore, the company that provides the paper towels could enter into replenishment contracts with buildings—charging not for product but for the number of hand-wiping events. This would enable a new business model: charge the company not for paper towels, but one dollar per building occupant per month. The product has been inverted to a solution that betters serves the needs of the user. That is one small example of the utility of a connectivity protocol with which most non-engineers are unfamiliar.

3. Defining a Suitable API

A device's network presence is defined by its API—the operations and data it exposes to the outside. This makes the design of the API crucial.

Whether accessing the device remotely or composing it with other devices, the API of the device determines what is possible.

At the IoT stage, a device just carries out the specific tasks envisioned by the designer, and this is reflected directly in its API. In many cases, there is a one-to-one correspondence between the API methods and the physical controls on the device. As an example, for the coffeemaker, there is an API method to start brewing (such as the start button), to stop brewing (ditto), to see whether the filter needs changing (a light), and to indicate whether brewing is finished (another light).

In designing a suitable API at the IoT stage, a good approach is to consider all of the available functions and then create an API method corresponding to each one. The idea is to give a user who might be interested in remotely controlling the device—or composing it with others—the ability to do *anything* via the API that he could do by physically touching the device. That means *nothing* should be left out.

Meeting New Needs

Despite concerns over security and privacy as discussed in detail in chapter 3, the potential benefits of IoT far outweigh its drawbacks. It is now possible to make new things and meet new needs merely by combining things that already exist in new ways. This opens up new realms of possibility. A company can expand into meeting new needs not by introducing new products, but merely by finding new ways to connect its existing products. Consumers themselves can do the connecting, turning them from passive experiencers of the company's vision into active partners in the design process. The principles of inversion become a tool for consumers as well. Whether they realize it or not, they become designers themselves, because they now have the flexibility and composability to meet their own needs.

However, exciting as IoT is, it's important not to fall prey to "technology push," a phenomenon that's been around since the dawn of engineering. Engineers will find some new thing they can do and push products out there that might seem cool but have no real purpose. With

IoT, everyone seems to be saying, "We've got to put the Internet in everything. Let's put it on this. What else can we do with it?"

The inversion paradigm defines a clear business purpose against which any IoT solution can be tested for relevance. It's not enough to put a device on the Internet for its own sake; it must advance a customer need in alignment with the needs-based mission of the company. Remote monitoring and control can meet customer needs by giving the device manufacturer a clearer picture of those needs and how well they're being met, suggesting new ways the company can take action to fill those needs. Giving remote access directly to the customer can empower the customer to find novel ways to meet his own needs—and even create new solutions by making new connections between devices.

But IoT is about much more than meeting needs through connectedness and composability. As powerful as that combination may be, there is more to come. While IoT enables remote access to things and the creation of new things by composing old things through intelligent mediators, an individual thing in isolation is still limited to performing the one task its designer envisioned. This is because the computer inside, no matter how powerful, is still running a fixed program. It is still mired in the quicksand of firmware.

The next stage occurs when we break the barrier of firmware and treat the computer inside the device as a full-blown computer that can run software. We call this stage the Intelligence of Things.

5 The Intelligence of Things

Breaking the Boundaries of Function

In 1990, Nippon Telegraph and Telephone Company released a new mobile phone the size of a man's palm and weighing half a pound. Able to maintain a phone call for up to forty-five minutes, it was a milestone in terms of size, weight, and battery life (if more like a millstone to carry). However, it still looked very much like the land line in your house—only smaller. At the time, mobile phone design was wed to the form factor of the traditional phone: it had a dial pad and a place for your ear and your mouth. Nothing suggested to the user that there was a computer inside the phone. It was controlled by firmware, and like the coffeemaker we discussed in the previous chapter, it was designed to do a single thing: make phone calls.

Over time, phone designers figured out a few other things a mobile phone could do. They added text messaging (awkwardly at first via the same ten-button keypad, but eventually heavy texters could get phones with a tiny QWERTY keyboard). With the release of the Nokia 7110 and its wireless application protocol browser, designers added extremely limited web browsing, but the companies limited access in the hope of one day becoming gatekeepers for this content. In 2000, Sharp debuted the first camera phone in Japan, and now phones could take pictures.[1] But each time such functions were added, it was just another bell or whistle: the phone could still only do exactly what the designer intended, even if that list of features was now slightly longer. Even the BlackBerry, whose QWERTY keyboard and mobile e-mail functionality helped make it extremely successful, was still basically a phone.

Then in 2007, Apple debuted the iPhone to great fanfare and equally great skepticism. No wonder: it was a radical reimagining of the phone.

It did the same things other phones did: made calls, sent and received text messages, and took pictures. But how it did this was revolutionary. Instead of these functions being built into the firmware of the device, Apple treated the iPhone's microprocessor as a full-blown computer, something on which could run any number of different software applications. All of a sudden, the phone was no longer limited to doing what a phone did. It could do anything that a software designer could imagine.

The user experience of the iPhone is that it can potentially do millions of different things. The user simply downloads an app and just like that, the phone has a new feature. Unlike all previous iterations of the mobile phone, the iPhone's features are not fixed in stone on the date the phone is released to the market and the user manual printed. New apps are created and made available every day, designed to perform a huge range of tasks from the ridiculous to the sublime. You can use it to summon a taxi, as a level for straightening a picture, as a flashlight, as a virtual reality headset, or anything else a software developer can dream up.

Computers Can Do Anything

The iPhone shares a key characteristic with the desktop PC: its functions are limitless.[2] Because it's a computer, its only restriction is the ability to run the software applications its user downloads to it. Anything you can imagine a computer doing, you have the power to make happen just by writing code. It is intoxicating. You don't even have to be a software engineer to experience this reality. One of the defining characteristics of this stage is that it's still possible for non-engineers to appreciate the limitlessness of a computer's function. No one is surprised if one day their computer can play movies, play music, or make international video calls. Because we grasp the power of software, we know that the computer we buy today will be capable of things tomorrow that we didn't know about when we bought it.

The iPhone and the smartphones that followed it transformed the mobile phone in the same way. Instead of a phone-like device running firmware that only lets it perform a few functions that the designers built into it, the smartphone is a computer running software. Software expands the things a phone can do toward infinity in the same way it expanded the functionality of desktop and laptop computers. Now there are thousands of new smartphone apps every day, so every day, your phone can do more stuff. Users are no longer surprised when their phones can give them the fastest route to their destination, tell them what commercial flights are overhead, photograph wine bottle labels and retrieve reviews and tasting notes, or help them invest their spare change in the stock market.

In the last chapter, we drew the connection between IoT and inversion. Early IoT devices can allow businesses to meet customer needs in new and innovative ways, but the limitations of their firmware mean that they are fundamentally inflexible. Once a device of this sort ships, its future utility is constrained by its inability to run software. Designers or engineers who have a high-benefit idea for delivering new value to their customers may be able to upgrade *existing* functionality via firmware update, but they may be unable to make their products do something *new*.

In this chapter, we now move beyond the connectedness of IoT and enter the Intelligence of Things, where function is no longer limited by form and design but is flexible, adaptable, and *recruitable*. In the inversion model, this is the stage at which businesses gain not only the power to quickly, cost-effectively, and proactively launch, test, and revise new solutions, but also to engage the customer as a collaborator in finding new uses for its solutions.

Recruitability

The defining characteristic of the Intelligence of Things is recruitability. As we observed in chapter 3, recruitability is one of the components of the inversion vocabulary. Recruitability is the ability of software to recruit the hardware components of a device such as the iPhone—its sensors, actuators, computer, or connectivity—to do new things that the designer did not originally build into the device. It enables an inverted business to respond rapidly to new customer needs not by designing and building new products but by using software to recruit the hardware from existing products to meet those needs. From a user's point of view, a device with recruitability is always doing new things. In fact, one of the hallmarks of the Intelligence of Things is that we are no longer surprised when devices such as the iPhone can do something new.

Desktop computers and smartphones are today's most familiar recruitable devices, but there are other examples:

- *The Amazon Echo.* We've written about "Alexa," Amazon's voice-controlled home assistant that allows users to order products, request services such as traffic reports, and control a home's connected devices with a voice command (what Amazon calls "skills"). Now, the Alexa Skills Marketplace (an app store for the Echo introduced in November 2016) lets users choose from tens of thousands of applications that add new functions to the Echo through an implicit process of machine learning. By clicking "Enable" and downloading an app to their Echo unit, users can ask Alexa to read their child a personalized bedtime story; have the sensors in their smart mattress begin monitoring their night's sleep; text, e-mail or call friends, family, or caretakers in case of an emergency; and a lot more. The Alexa Skills Marketplace allows users to recruit the Echo's cloud connectivity and processing power to meet an increasing range of needs, from the whimsical to the critical.
- *The Nest Learning Thermostat.* We have written about the Nest as an example of the remote sensing and remote control capabilities characteristic of IoT. However, the company's Works with Nest protocol allows a Nest device—not only a thermostat but a Nest Protect smoke alarm or Nest Cam home security monitor—to seamlessly connect with and control other connected devices in a home. For example, when a user with a Nest Learning Thermostat adds a smart light fixture to her home's wireless network, the Nest automatically detects the new device. Using the app for the light fixture, the user can connect it to the Nest. Now, if she leaves the house without turning off the light, the Nest recruits its sensors to determine that no one is in the home and an actuator to turn off the light. With an increasing number of IoT devices bearing the "Works with Nest" symbol, the company is aggressively embracing an inversion paradigm—selling not temperature control but comfort, security, and convenience.

- *Tesla Motors*. In the past, if a consumer wanted to add a new function to his car, he faced a time-consuming, expensive, and often frustrating process of installing new hardware—and also being without his vehicle while the work was completed. Tesla has disrupted this paradigm by upgrading existing functions and adding new functionality to drivers' vehicles using software updates. A scheduled update might not only modify a car's Autosteer function, but also let owners link driver profiles to specific key fobs—recruiting the vehicle's sensors to determine an approaching driver's identity and its actuators to adjust seat and steering wheel positions.

Other software applications that leverage recruitability to enable new functions include: the "creep" setting, which engages the electric motor to roll the car forward slowly when the driver removes his foot from the brake, mimicking a gasoline-powered car; Traffic-Aware Cruise Control, which recruits the vehicle's radar, cameras, and ultrasonic sensors to speed up or slow down based on the traffic around it; and a 2013 software update that recruited the Active Air suspension systems in some cars to change the suspension settings and allow more clearance at high speeds.

In *Wired* magazine, writer Alex Brisbourne perfectly summarizes Tesla's place as an exemplar of the Intelligence of Things: "In nearly all instances, the main job of the IoT—the reason it ever came to be—is to facilitate removal of non-value-add activity from the course of daily life, whether at work or in private. In the case of Tesla, this role is clear. Rather than having the tiresome task of an unplanned trip to the dealer put upon them, Tesla owners can go about their day while the car 'fixes itself.'"[3]

The Three Architectural Keys to Recruitability

The defining characteristic of recruitability is that it frees a device from being limited to the functions it was originally designed with, using software to enable its hardware to perform new functions and meet new needs that the designer may never have imagined. Because of this, recruitability is essential to the inversion paradigm, so as the Intelligence of Things becomes more widespread and more businesses shift to an inverted business model, designers need to know the architectural factors involved in making devices fully recruitable.

First, the computer at the core has to be powerful enough to host new applications. It must have sufficient processing speed, have a large amount of RAM, and large quantities of nonvolatile memory for permanent storage of applications and their data. For example, the 2.3 gigahertz multi-core A10 processor in an iPhone 7 Plus is 120 times faster than the original iPhone and the phone has three gigabytes of RAM and up to 256 gigabytes of nonvolatile storage—specifications comparable to those of a low-end desktop computer. Putting this kind of power into a device is a completely different mentality from the firmware mentality of IoT. The firmware mentality focuses on reducing cost by using the least powerful processor and the least amount of memory needed to do the job. But the designer doesn't know what applications will eventually run on a recruitable device, and the choices of processor and memory will ultimately limit what is possible. For recruitability, designers must build devices with the most powerful processor and the most memory that can be achieved for a reasonable cost—not the least cost.

Second, there must be a way for the device to host new applications, which implies two other essentials:

- A way for application authors to create those applications and make them available for users.
- A way for users to discover those applications and run them.

For a recruitable product, designing this aspect is just as important as designing the inside of the device itself. Apple understood this when they built the iPhone, and they made two key innovations in this area. First, they created the App Store, which made the process of finding, purchasing, downloading, and installing a software application extremely easy—far easier than it was (and still is) to put a new software application on a personal computer. Instead of using the installation wizards customary with the PC, Apple was forced to make this process easy because of the limitations of the small phone screen. However, from the point of view of recruitability it was a masterstroke, because the ease of download and installing new software encouraged users to constantly recruit their devices to new tasks.

Apple simultaneously created a business model for app developers that made it very easy for an application developer to get his application out into the world and make money from it. This was a step change from the PC software world, in which for years distributing new software meant creating an installation program, putting it on distributable media such as a CD-ROM or creating your own download site, advertising it, and providing customer support when the inevitable problems occurred. If you also wanted to make money from the software, you also had to create an entire sales and payment infrastructure.

But with the App Store, any developer could instantly make a new application available to the entire iPhone user base with a single click. If the developer wanted to make money, she named her price and Apple would give her a generous share of the revenue from each download. By doing this, Apple created incentives for millions of software engineers to make new content for the iPhone. Each time a new app was released, giving users the ability to recruit the iPhone's hardware to do new and amazing things, the iPhone became more useful to its users—and the paid apps also earned Apple new revenue.

This shows how recruitment works in multiple directions to enhance value. Each new app recruited the iPhone to a new purpose, which satisfied the customer needs that the app developer sought to fulfill and created more value for iPhone owners. At the same time, the need to make the iPhone more recruitable through apps led Apple to create its incentive-laden ecosystem, leading to a wealth of new apps and greater value for app developers.

The Apple App Store or Amazon's Alexa Skills Marketplace are public-facing solutions for offering recruitability. In other cases, recruitability might be more invisible. For example, the Tesla Model S is recruitable, because Tesla Motors can recruit the vehicle to new tasks simply by remotely downloading software. The users have the pleasant experience of waking up to find their car can do new things, thanks to the previous night's download. However, the Tesla is a more controlled form of recruitability than the iPhone, because only Tesla can create new apps for obvious and sensible reasons of safety and government certification.

The third thing designers must do to build recruitability into devices is to generalize the hardware. Recall that an IoT device consists of sensors, actuators, computation, and communication. Just as the computation component has to be enhanced to support recruitability, a device can be made more recruitable by enhancing its capabilities to sense, actuate, and communicate. Just a decade ago this would be cost prohibitive because the components were expensive, but today adding a multitude of sensors, actuators, and communication methods is cheap. Some of the things that designers should consider adding are:

- *Sensors*
  - Accelerometers (tells you if the device moves and how much)
  - Orientation (measures the device's orientation on the compass and which way the user is holding it)
  - Location, including GPS, cellular triangulation, and various indoor location methods such as Wi-Fi triangulation, beacons, and so on
  - Image sensors (cameras)
  - Microphones
  - Touch, including a touchscreen
  - Buttons and switches
  - Temperature/humidity
  - Battery level
  - Radar
  - Lidar
  - RFID
- *Actuators*
  - Loudspeakers/headphones
  - Haptics (controls that vibrate, creating the illusion of tactility)
  - Switches
  - Heating elements
  - Lights and other fixed-purpose displays
  - Screens (including touchscreens)
- *Communication*
  - Wi-Fi
  - Cellular data (3G, 4G, soon to be 5G)
  - Bluetooth

- ZigBee
- 6LowPAN
- LPWAN

For recruitability, the key is to put in as many of these sensors, actuators, and communication methods as can be done at a reasonable cost, as the designer does not know in advance to what purposes they will be recruited. One example is the Sky View iPhone app, which uses the phone's orientation sensor to calculate which part of the sky your phone is pointing at, then superimposes maps and information about celestial bodies (for example, the names of planets or stars) over the view of the sky. It's doubtful that the designer of that sensor anticipated that it would be used in such a way—until a creative software engineer saw its potential.

Also, with recruitability there should be a bias toward building in more general-purpose sensors and actuators. The iPhone, for example, has far fewer buttons and indicator lights than the cell phone it replaced, yet it can do more because the touchscreen is far more versatile. As a general design principle, if you can include a hardware component without making the device too cumbersome or costly, do it. The designers of the Nest Thermostat included an 802.15.4 (ZigBee) transceiver and the proprietary Weave connectivity protocol in its early models, even though they didn't have a use for it yet. Then, when the company launched its Protect smoke/carbon monoxide detector, the two devices could talk to each other without using the home's wireless network.

When in doubt, add the hardware.

The Power of the API

In the last chapter, we identified the API as one of the foundational technologies of IoT because it makes a device's internal functionality available to the outside user. The design of the API for a recruitable device is still crucial, but now that API not only exposes functions for external integration, but also for applications hosted directly on the device itself.

For example, the internal APIs provided for iPhone and Android apps are rich, powerful tools. Where an IoT device's API might expose a few high-level operations that express the intended purpose of the device, the API for a recruitable device exposes each of the sensors, actuators, and communication devices separately. The logic that carries out the original purpose of the device is not firmware, but is just another app that uses the same API that other app developers will use. This makes individual hardware components, from accelerometers to switches, accessible and recruitable for creating new functions.

Consider, for example, our familiar coffeemaker reimagined as a recruitable device for the Intelligence of Things. Like the IoT coffeemaker from chapter 4, it has sensors (temperature, water level), actuators (heating elements, water valve), and communication. But instead of inflexible firmware, it now has a powerful embedded computer that is capable of hosting apps. The logic for brewing coffee, formerly burned into the firmware, is now just one of many apps the coffeemaker can host. Like the telephone app on a smartphone, the coffee brewing app for the coffeemaker is provided by the manufacturer and preinstalled onto the device when the user unpacks it, but under the hood it is just another app.

Now suppose that some coffee connoisseur discovers that her particular style of coffee is best brewed at a different temperature, or she wants to have the first half of the water delivered at one temperature and the remaining half at a different temperature. No problem—the company that builds the coffeemaker need only write a new app that controls the temperature sensors and the heating elements with different logic. Or if the connoisseur is herself a software engineer, she might be the one to write the app.

As flexible as this reimagined coffeemaker is, it becomes even more so if the designer has designed it for recruitability and added more powerful sensors and actuators in addition to the temperature sensor and heating element. If the coffeemaker has a motion sensor, a revised coffee brewing app can immediately shut the coffeemaker off if it detects the coffeemaker has been accidentally tipped over. If the company adds a loudspeaker, the coffeemaker can now shout out "Coffee's ready!" when brewing is complete—and if the company has also added

Bluetooth connectivity, it can pick up the signal from the Pandora app running on the user's iPhone and fill the kitchen with morning music.

If the coffeemaker is given a touchscreen instead of a simple start button, it can show a picture of the different kinds of coffee to be brewed . . . but more recruitably, app writers can now create all sorts of apps to transform this simple appliance. There might be an app that keeps statistics on how much coffee has been brewed or shows you a graph on the screen of how your caffeine consumption is changing over time. Or another app that keeps track of the favorite brews for each family member. Or one that reaches out over the Internet to download and display a weather forecast that you can peruse as you pour a cup from the pot. Or something that talks to your blood-pressure cuff and reduces the caffeine level in the morning's brew. Or orders replacement pods. Or brews tea. Or hot chocolate. The list is endless . . . just as it is for a smartphone.

Three Types of Recruitability

The power of the API highlights the reality that there is not one type of recruitability, but three. The first is the most obvious, *designer recruitability*. The designer of a device and the business that manufactures and sells it creates the device with hardware that can be recruited to perform new functions and an API that exposes that hardware to its own subsequent software apps. After the device has been sold and is in use, the business can release software that allows the hardware to be used in new ways, as in the Nest Thermostat's transceiver.

The second type of recruitability is *third-party recruitability*, and it is this type that truly taps into the inversion potential of the Intelligence of Things. With third-party recruitability, businesses and individuals other than the company that originally designed and sold the device also create software that recruits its hardware to do new things and meet new needs. That's the innovation model that has made the iPhone and its App Store such a powerful combination; by establishing a platform and then creating an ecosystem that allows others to leverage it, you expand functionality and create value for multiple stakeholders.

Third-party recruitability also works in the other direction. If you are a company working toward an inversion model, perhaps you have identified a customer need that you can meet but are reluctant to invest the time and capital to build your own connected device. Now you can partner with another business selling connected, recruitable devices to some of the same people who are also your customers and develop a software application that will recruit your partner's hardware to deliver new services to your customers! That model has turned thousands of individual app programmers into successful small businesses thanks to the iPhone App Store.

The third type of recruitability is *user recruitability*. Now, for all practical purposes, most of what appears to be users recruiting a device's hardware is actually composability—using a smartphone app to combine the existing hardware functions of multiple devices to meet needs. But the defining trait of recruitability is that it uses sensors, actuators, and the rest to do something *new* that the designers did not intend. Can users who are not software developers really do that?

Yes, if the device manufacturer permits it. If the coffeemaker's API only gives the consumer the ability to brew coffee and stop brewing coffee from his smartphone app, then all you really have is coarse-grain composability of the kind we discussed in chapter 4. However, if the API gives individual access to the coffeemaker's sensors and actuators, then a user can create new functions by composing those sensors in actuators in new ways. Users don't have to be software engineering experts because of easy-to-use composability frameworks such as IFTTT that work directly through APIs.

Should one type of recruitability be made available to all users? All three? This is one of many decisions businesses and designers will face as they embrace inversion.

Security and Privacy in the Intelligence of Things

This increased access to a device's hardware, which brings unimagined new functionality and value, also brings with it the potential for additional vulnerability in the areas of security and privacy. It is essential that designers ensure that the hardware components in the devices they build cannot be taken over by unauthorized parties, especially if they are opening the device's API to third-party apps. It is not difficult to imagine a bad actor misusing a device's sensors to spy on the user or using actuators to cause malicious harm in the environment.

Take the case of the four-star Seehotel Jaegerwirt hotel in Austria. Hackers penetrated the hotel's electronic key system, leaving the hotel unable to make new room keys for its guests until it paid a Bitcoin ransom equal to about $1,600.[4] That's an example of recruiting a system's actuators—in this case, the transmitters that encoded the passive RFID tags in card keys with new sequences of lock codes—and preventing them from performing their task. Security breaches could also result in misuse of the computational platform and communications hardware. These could be recruited in tandem to carry out denial of service attacks or infect other systems with malware.

Software security by itself may not be sufficient to protect against malicious activity. Security experts recommend hardware-level security measures: encryption of configuration bit streams that prevent unauthorized users from accessing a device's API and hardware functionality; tamper protection; secure key storage; and zeroization, the practice of automatically deleting sensitive cryptographic data if a piece of equipment is compromised. With many electronic devices now built by third-party contractors, often overseas, companies must also pay close attention to supply chain security to ensure their hardware is not cloned.

The abundance of sensors, actuators, computation, and communication in IoT devices also presents privacy challenges. Because privacy is largely a matter of intentional behavior by businesses, its protection hinges on being cognizant not merely of the hardware's potential to be recruited to capture or reveal PII but of its potential to literally invade privacy. The unauthorized operation of a device's video camera or microphone in a customer's bedroom, for example, could have obvious privacy implications.

Privacy questions also arise because of the power of the Intelligence of Things to collect more and more granular data on customer preferences.

Security: Mandatory for Self-Driving Vehicles

Few things bring home the potential vulnerability to hackers—or the importance of securing connected devices against intrusion—more than someone hacking into a self-driving car. That's precisely what Jonathan Petit, principal scientist at Security Innovation, showed he could do with a $60 system made up of low-power laser and a pulse generator. In a paper presented at the Black Hat Europe security conference in November 2014, Petit described how easy it is to trick an autonomous vehicle's light detection and ranging system into thinking an obstacle is directly ahead or to "spoof" it with so many signals that the car simply freezes in place. Given the safety implications of self-driving vehicles that can be hacked by malicious actors, it is unlikely that the autonomous car will achieve widespread adoption until this security problem is addressed.[5]

If you collect data about when people brew their coffee and what kind of coffee they like, over time you can assemble a substantial database on that individual that could be used for innocuous purposes—sending personalized marketing messages and electronic discount coupons, for example—or more ominous purposes, such as using brew time data to infer when the user is likely to be away from the home, making the home a target for a break-in.

The issues and solutions surrounding security and privacy in IoT are too dynamic and complex to be covered comprehensively in this book, but as we move into the Intelligence of Things and beyond, every business that wishes to pursue the inversion model should be aware of them and plan to incorporate them into its business plan.

Borrowed Recruitability

Business owners and product designers worrying that they will need to go back and redesign all their products to make them fully recruitable can relax. While recruitability is the most important characteristic of devices in the Intelligence of Things, it is not necessary for every device to have a powerful embedded computer and an app store in order to be recruitable because of the power of one of our other new engineering principles, composability.

The key to this is the smartphone. Just about everyone carries a smartphone these days, and the smartphone's touchscreen, camera, and audio capabilities can be recruited to enhance the capabilities of a device that lacks those things. Moreover, the smartphone has an app store and can communicate with other devices. The smartphone can bestow Intelligence of Things status on any other device it can communicate with—in essence, allowing a device that lacks full recruitability to "borrow" the smartphone's recruitability. Instead of outfitting our ubiquitous home coffeemaker with a touchscreen, the owner could use his smartphone to establish a Bluetooth link to it. The smartphone would then supply the touchscreen, Wi-Fi connection, and the processor and memory to host all sorts of coffeemaking apps. As long as the coffeemaker is in contact with the smartphone, the coffeemaker has the limitless potential of the Intelligence of Things.

Borrowing does require some change in the way an IoT device is designed. Specifically, its API must be designed for recruitability. A coffeemaker whose API only provides methods to start and stop brewing is not fully recruitable. In contrast, the recruitable Intelligence of Things coffeemaker exposes an API that allows individual control of the temperature sensor, the heating element, and so forth. This is the same API that a coffeemaker would have if it actually had an onboard platform for hosting apps. By combining recruitability with composability, that same degree of control is available using external software (such as an app running on a nearby smartphone), and in that way the device can do more than its own hardware would seem to allow, because its hardware can be fully recruited via its API.

There are tradeoffs involved. For example, if a coffeemaker recruits its owner's smartphone to host apps and provide a sophisticated display, it can only be used in those ways if the user has his smartphone handy. Also, borrowed recruitability adds an additional processing and communications layer that might render some functions impractical. For some applications, there may be a real-time constraint in how fast an actuator must be activated in response to a sensor—a constraint that might not be possible to meet if it requires a back-and-forth exchange over a communication network. That is yet another design consideration that a business operating in the inversion paradigm should be prepared to confront. Recruitability should be near the top of the list of design criteria to be prioritized with other goals and constraints.

These examples show how the smartphone has become a sort of "universal Intelligence of Things" remote, because it is ubiquitous and can be recruited to provide all of the qualities required to give the illusion that other devices have intelligence. However, as more types of devices also have native intelligence, the role of the smartphone may not remain as central in the future as it is today.

The Final Stage

We are currently in the early stages of the Intelligence of Things era, with just a few devices that are fully recruitable. But we expect that more will appear on the market in the near future. In this chapter, we believe we have shown that the arrival of such devices is not only inevitable, but also will presage a dramatic increase in the utility of IoT and of inverted businesses. As more and more designers and engineers create products that can run software applications and build APIs that are accessible to the end user, a nearly limitless ecosystem of intended and unintended functions will arise throughout the economy. Some will be novelties, but others will spark new products, business lines, and businesses for which inversion is the core mission.

In other words, with the Intelligence of Things, inversion will cease to be a reactive or accidental consequence of start-ups seeking a seam in the market or large companies trying to keep up with more agile competitors. Inversion will simply become how business is done.

That opens the door for the next stage: the Immersion of Things.

6 The Immersion of Things

When Devices Become Experiences

We began this book with a speculative scenario from 2030: a self-driving car as an important character in a story of a day's immersive activity. In it, the human at the center transitioned from home to car to office to car and back to home again seamlessly, the ordinary tasks of her day carried out autonomously, her environment morphing around her to accommodate her activities, preferences, even mood. We also told you that such a day was not science fiction. In this chapter, we explain why.

Today, the worlds of business and technology are experiencing the beginnings of a convergence. Technologies that were developed and deployed under different circumstances and for widely varied purposes are starting to blend together as the Intelligence of Things becomes more and more common in homes, businesses, hospitals, airports—everywhere. Over the next few years, the connected devices of IoT will converge with the augmented reality of Microsoft's HoloLens. Machine learning will converge with robotics. Together, all these technologies, architected by expert engineers, will create a seamless "hyper-reality" in which customers are not even consciously aware of interacting with "devices," but instead move through the world naturally as the surrounding environment meets their needs.

This may seem like overheated prose for talking about fanciful concepts, but this converging reality already exists in small pockets—the self-contained, responsive ecosystem of the Tesla Model S autonomous vehicle, for example. It is growing daily. Inevitably, the lines between devices and technologies will blur as things interact with things invisibly, recruiting functions, and collaboratively delivering benefits in such natural, organic ways that we ask, "Was that the self-driving car or my home-based artificial intelligence that made the appointment for my suspension tune-up . . . or both?"

When users don't know—and don't care—whether value comes from one system or all systems working in concert, we have arrived at the "all of the above" world that we call the Immersion of Things.

What Is Immersion?

As things become more intelligent—as sensors and actuators, and device functionality become more amorphous, as wireless communication becomes faster, cheaper, and even more ubiquitous than it is today—they will form a pervasive sea of resources that can be recruited beyond the functions of a single machine. The apps need not be limited to an intelligent thing, and can, instead, span things and the environment. We will live immersed in this sea.

Just as sensing will be everywhere, so will computing. Real-time control, where activity requires the kind of immediacy that can't be achieved with a long roundtrip to the cloud, will occur via an array or "mesh" of end-user clients or near-user devices that reside at the edge of the network rather than at the center. When deeper computing is necessary, processing will make side trips to a centralized private, edge, or public cloud. This is what, roughly speaking, Cisco refers to as "fog computing."

A cloud floats overhead, its very distance implying remoteness. Fog exists at ground level, enveloping everything around it. That is the reality to come, when we move forcefully from first selling products, then to meeting needs, and finally to creating and curating experiences. Technology and business leaders in Germany have been talking about this stage, Industrie 4.0, since 2011, talking about cyber-physical systems and their impact not only on business and manufacturing but on government, society, and human identity.[1] Now, that stage is right around the corner.

This version of IoT enables a new form of augmented reality and makes for a future of extraordinary possibilities. It will draw on other technological fronts including machine learning, autonomy, and speech. It's exciting, and it's scary for many reasons and brings with it many challenges, not the least of which will be security and privacy and its impact on jobs and the workplace. However, businesses and innovators who rise to the challenges will usher in a new world.

The Immersion of Things goes beyond the Intelligence of Things because it is based on composing behaviors and experiences by recruiting from a sea of amorphous resources—sensors, actuators, and connectivity—that reside everywhere, some fixed, others mobile, some permanent, others transient. Consider TrackR and Tile, two companies whose value proposition depends on crowdsourced GPS. Both systems recruit the Bluetooth and GPS connectivity on participants' smartphones to connect to the cloud and locate lost objects—car keys, for instance—fitted with a small Bluetooth tag. Suppose you put a Tile tag in your purse. Whenever you are near the purse, your smartphone will read the Tile tag via Bluetooth and record where it was last seen. But suppose you accidentally left your purse somewhere in the shopping mall. As *other people's* smartphones pass by your purse, those phones will also record the location of the purse, securely, without the other phone owners' knowledge. The other shoppers' smartphones have been recruited to help you find your purse; once found, only your phone will read your purse's tag.

This is an extraordinary example of *participatory recruitment.* It's a dance that smartphone designers did not anticipate but, like a flash mob, are happy to participate in. The "recruitment" of the phone's components turns up in permissions that ask you to share location. That is a minor example of the sea of recruitable, collaborative functionality that is becoming commonplace as we move toward immersion.

In the previous two chapters of this section, we have discussed the inverted, needs-first business model from the perspective of devices—pieces of silicon, glass, plastic, and metal that perform specific functions and that can be composed to perform even more functions. Now, in this final chapter tracking the evolution of things, we move into an environment where the devices, for all practical purposes, disappear—or at least become irrelevant. The ability to anticipate customer needs and even predict those needs before the customer is aware of them, is the ultimate example of the inverted business model in action—of responsiveness and problem solving so intuitive that it appears to be supernatural.

In the first part of this chapter, we will take a deep dive into the complex web of technologies and engineering schema that make immersion possible, including the many protocols that businesses can use to share data. Then we will offer two examples—one from the consumer's experience, one from an industrial setting—that reveal what immersion might really look like.

Architectural Principles Powering Immersion

• *Discoverability*. The components of more and more devices will be *discoverable* by other devices for the greater good. Some will be publically accessible, while others will be privately accessible. For example, the parking meter of the future may also be a charger for an electric car. When the car pulls in, its charging capability will be advertised either over the cloud or using a local beacon. An onboard display will indicate the price per kilowatt-hour, and the cost of parking (free if you are charging). Once you park, you may be able to recruit a tiny camera in the parking meter to watch over your car.

There are two types of discovery: *static* and *dynamic*. Static requires a specific setup and handshake. Dynamic is fluid and ad hoc: your car talks to a new parking meter because it is set up to talk to any parking meter. The Immersion of Things will push dynamic discovery and ad hoc functionality.

Security will be conditional. A repair person may be able to access the internals of the parking meter—a temperature sensor, for example, to ensure that vehicle charging is not overheating the meter's circuitry. During an AMBER Alert to find an abducted child, the city may volunteer the cameras on all its meters—and ask you if it can also access the cameras in your vehicle—to collaborate in the search for a silver Hyundai Elantra with a particular license plate.

Auto accidents can also create these "all-in" situations. Usually, the antilock braking system in your car speaks only to your car. But when you hit black ice, there is no advantage to keeping that data secret. Your antilock braking system will issue a radiofrequency emergency message telling nearby cars to steer clear. Sensors in the road dividers will indicate if you have recovered from your skid or slid across lanes or even onto the shoulder. If the cloud detects damage, it will automatically contact towing or emergency services, while your phone's Waze app alerts preselected friends or family that you've had trouble on the road and maps your location.

Discoverability will recruit your device and its components into a larger experience. Be prepared to be a player in someone else's story.

- *Context*. In an AMBER Alert situation, when your car reports the license plates it sees on the road, that information is useless without context. Where were you when your camera detected a certain plate? In which direction were you heading? Any device that is discoverable needs to provide the context around the activities in which it is used. A charger on a parking meter must communicate the type of current and voltage it can support.

If you create solutions that will communicate information, think of the context needed to make that information relevant. Context has many faces, some of them very basic information that is easy to take for granted. The temperature sensor in a solar-powered road reflector may need to indicate not just temperature, but which lane it is in. Information that is adequately contextualized will typically include dimensions of *what, when, where,* and (most importantly) *why*. Consider a ski resort that uses RFID to track its rental equipment using RFID tags. A fully contextualized RFID event that describes a pair of skis being sensed would have all of these dimensions:

- *What*—the unique identifying number for the pair of skis, as read from the RFID tag.
- *When*—the date and time at which the skis were detected by the RFID reader.
- *Where*—the location where the skis were detected; for example, in the rental shop at the north slope lodge.
- *Why*—the business context of the observation; for example, the skis were being *rented* to a customer, are now in a rented state, and the rental contract number is 271828.

With contextualization, raw sensor data is enriched by the application that does the data capture. In the ski resort example, it's not enough for the RFID reader in the rental shop to report the raw tag read (which only includes the *what* and *when* dimensions); it must enrich the data to include the *where* and *why* information. Use of data standards for contextualized observation data can help in creating suitable data.

• *Orchestration*. Someone needs to build a system and determine its behaviors. The behaviors can be organic; others may need a dedicated system to orchestrate them. In the Tile example, the behaviors are organic. But at home your Amazon Echo, Apple TV, Nest Thermostat, and even your cable set-top box are all competing to own your behaviors. Walled gardens mean limited compatibility and ease of integration. Patchwork connections are possible—Echo and Nest can talk to each other, for example—but the overlapping rules and behaviors potentially breed chaos. Before long, a complex series of mistakes will result in a local news story about some poor homeowner being locked out of his house on a cold January night by his television.

For reasons of security and access control, and because organic workflows could create unintended consequences, immersion is best architected by having a logically centralized software system that can orchestrate all behaviors, but which delegates operations to the edge as needed to achieve real-time responsiveness. This is the architectural principle of *mediation* we saw in chapter 4, but in the Immersion of Things, it's even more critical because connections between devices are dynamic.

Is your business model built around orchestrating and owning of workflows, or providing a critical component for other businesses' workflows? These are key questions.

• *Recruiting non-digital objects*. In the Internet of Things and Intelligence of Things stages, we focused on digital objects that contained sensors, actuators, computation, and communication. But even as digital technology penetrates further into everyday objects, there will always be objects that simply won't ever host an onboard processor or a battery: a pair of socks, a box of tissues, a bottle of milk. But in a truly immersive environment, even these objects must be recruitable into an overall experience.

The trick is to make such objects recruitable by giving each a unique identifying number that lets it be sensed by some other device, such as a smartphone. When the sensing device detects the presence of the object, it communicates that information (contextualized), and applications can act as though they are interacting directly with the object. Automatic identification and data capture technology makes this possible, and there are varieties of it:

- *Passive RFID*. A tiny silicon chip containing a tiny amount of data: usually just a unique identifying number. The tag has no onboard source of power, but it does have an antenna that can receive a signal from a nearby reading device. The antenna extracts just enough power from the incoming radio signal to power up the chip and allow it to respond with the unique identifying number. RFID tags can be read from up to ten meters away and can be read even when there is no direct line of sight to the reading device.
- *Bar codes*. A pattern of dark and light bars or squares that can encode a unique identifying number. They can be read by specialized scanners as well as the digital cameras as found on smartphones. While bar codes do require a direct line of sight, they are extremely inexpensive as they can be printed by any ordinary printing device.
- *Digital watermarks*. Similar to a bar code, but instead of using visible light and dark bars, a digital watermark uses shades of color that are so subtle that the human eye does not notice their presence. An object such as a consumer product package can be covered in copies of the same digital watermark, allowing the object to be read regardless of orientation yet have no human-visible markings.

All of these automatic identification and data capture technologies make it possible to recruit even very inexpensive, disposable everyday objects into immersive experiences. For example, Tile allows users to outfit their phones, car keys, or any other small objects with RFID tags so the objects can easily be located using their smartphones. No more locating lost keys while trying to leave for work; now, they simply appear on your phone's display.

- *Machine learning/deep learning*. Immersion is about making the consumer's environment instantly respond to his needs without his taking explicit action—ideally before the consumer is even conscious of the need. This is possible thanks to advances in machine learning. Though the term "artificial intelligence" is sometimes applied to this technology, it is really just a convincing illusion of intelligence.

Machine learning simply recognizes patterns in data and predicts an appropriate response to new data based on what worked in the past. IoT technology makes it possible to generate enormous quantities of training data, while advances in computing have made it possible to create "deep learning" networks that can digest such quantities of data and thereby achieve uncanny precision in responding to patterns.

For example, an earlier generation of machine translation tried to use human-assembled dictionaries and parts of speech, which often yielded comical results. Deep learning systems digest millions of pages of popular books and government documents in multiple languages, and as a result can translate nearly as fluently as an expert human speaker. In the Immersion of Things, deep learning can digest millions of observations of users interacting with the world and use that to anticipate needs and respond without the user having to ask.

- *Collaboration.* While it is possible to create a fully immersive environment that is entirely under the control of a single business (e.g., the interior of a Tesla car), it is far more likely that companies will collaborate to create immersive experiences. For example, a business traveler in a truly immersive world will be able to initiate a videoconference on his smartphone as he walks from the airport gate to the exit (with an inset in the corner of his screen helping him navigate the airport's floor plan), continue the conference on the screen in the backseat of his Lyft car, and have it seamlessly transfer to the large screen in his hotel room as he walks in the door (his smartphone having already checked in and obtained the necessary credentials to unlock the door by Bluetooth).

This scenario can only happen through orchestration, with the travel app provider, the airport, the ride service, and the hotel all collaborating behind the scenes. Businesses must be prepared to share data as never before, not merely execute transactions, as their customers move from one place to another. (See the following Tech Alert! box on collaboration technologies.)

Tech Alert!

Data-Sharing Architectures

There many ways that businesses can share data across company boundaries:

- *M2M messaging*. Two businesses that establish a collaborative relationship can send messages directly to each other. A common example of this is electronic purchasing and invoicing, commonly called Electronic Data Interchange. Each party keeps its own data and shares data only as necessary to achieve a given business goal. It gives great control to each party, but it can make it difficult to ensure that both parties have current data, or to share data between parties who are not direct trading partners and lack a preexisting M2M communication path.
- *Trusted, centralized third party*. Businesses can collaborate by using a trusted third party to store shared data. This is simpler than M2M when many parties are involved; instead of connecting to each peer, the company has a single connection to the trusted third party, which mediates data exchange between the parties. Finding relevant data is easier, too, as it is all in one place. Trusted parties are often used when there exists a party that all parties consider neutral, for example, a government-operated service.
- *Federated*. In the federated architecture, each business shares its data with a third party that delivers the data to the right peer. But instead of there being a single third party, there are many to choose from. This avoids all players having to agree to a single third party, but it adds complexity when the sender has chosen one service provider and the receiver another. This often requires a neutral "root" service to coordinate the service providers. One example is the Internet Domain Name System, where individual domain name records are held by different registries that are federated by a root service governed by an international organization.
- *Decentralized ("shared ledger" or "blockchain")*. In a decentralized system, there are again multiple service providers (as in the federated architecture), but each has a copy of all the data. Each time a business needs to share a new piece of data, it sends it to one of the service providers, which then copies it to all others. The providers use a consensus-generating algorithm to ensure that they all have a consistent copy of the data and that no one acting alone can censor or tamper with the data.

The most mature system of this kind today is the Bitcoin network, where the shared data are the records of transfers of virtual currency

Tech Alert! (continued)

between parties. The inventors of Bitcoin were explicitly trying to create a means to transfer value that could not be interrupted or subverted by any government or central bank. So the governance property of the decentralized system is paramount. Shared ledgers are also promising for complex, multi-party financial transactions, where it is important not only that "I know" and "you know," but also that "I know that you know."[2]

It remains to be seen what role if any decentralized systems will play in IoT applications, but to the extent that consumers seek to avoid having data about their personal objects not be under the control of any one business, the decentralized governance of shared ledger systems may prove important. To the extent that consumers generate large amounts of data that can be mined for deep learning, shared ledgers offer the promise of greater transparency so that consumers know who is using their data, and perhaps can be compensated for its use.

On the other hand, this technology is still in its infancy, and in particular it remains to be seen whether shared ledgers can scale to the data volumes that can be expected from IoT applications.

Mama's Got a Brand-New Bag

How might immersion and its foundational principles play out in the real world? Here is an example from the consumer products realm.

It's eight o'clock on a sunny March morning in San Francisco in 2019. As an executive leaves home, she grabs her "smart purse." The purse is equipped with a small, rechargeable battery, motion sensor, accelerometer, outside-facing camera, computer, GPS unit, and a software-defined radio that speaks RFID, Bluetooth, ZigBee, cellular, and a variety of other wireless communication protocols. As the woman leaves her apartment, the bag's RFID reader detects motion and sees the RFID tags on her glasses and her issue of a technical journal go out of range. It emits a tone to remind her that she has forgotten these items. She reaches back and picks up her glasses but the journal, alas, is an old, outdated issue.

The executive is making a day trip to Los Angeles, and during the ride to the airport, the bag notices that her phone is low on power and charges it. Inside the terminal, she passes by a shop with a Bluetooth beacon that sends her smartphone a reminder to buy a new issue of the tech journal, and she stops in and buys a copy. Her boarding pass resides on her phone, which updates gate information and time until boarding and vibrates to let her know when she should walk to her gate.

As she approaches the gate area, the bag notices that its internal battery is low and looks for chargers advertising their services over Bluetooth. It discovers one and sends a message suggesting that the woman sit near it, where it charges her battery automatically while paying for the service electronically after she uses her watch to approve the transaction. After a few minutes, the woman needs to use the restroom and leaves her bag on the seat next to her, taking only her phone. However, her bag is perfectly safe: the bag notices that her phone is out of range and locks itself. If the accelerometer detects motion, it will set off an alarm and message her on her phone. Last week, she also downloaded an app that registered the bag with airport security, so it someone tries to tamper with or steal it, the bag will let out a radiofrequency "scream" to alert nearby security cameras that there's trouble—but also to reassure security that the bag belongs to a secure traveler.

Arriving in Los Angeles, the executive leaves her bag in the plane by mistake. Fortunately, she is able to find it using crowdsourced GPS. As people file past her phone in the plane, their phones read the bag's beacon, telling her through the cloud that it's lost and giving her its location. She retrieves it from a flight attendant and goes on with her day.[3]

Inversion and Sports

Sports will likely track the entire evolution of IoT. Imagine playing tennis: already, the Babolat Play PureDrive tennis racket comes with a chip in the handle that contains a three-axis accelerometer and a three-axis gyro. Using built-in sensors, it can measure racket head speed and how much spin is imparted to the ball. In the future, the racket might recruit 3D location beacons to tell the player how many times she went to the net and whether she took a stutter step en route. Placing sensors on the ball will be difficult, but if someone cracks that problem, imagine the implications for coaching placement and power. This use of technology reinvents how a customer base meets its needs.

The implications for team sports are also extraordinary. Zebra Technologies has teamed up with the National Football League to track players on the field to within six inches. Ultra-wide-band tags are placed in the shoulder pads of the players and readers are placed around the stadium. Data including the location and the speed of each player are sent in real time to commentators so they can use the information during the broadcast to increase viewer engagement. Coaches can also use this information to figure out their game plans, get aggregate statistics on a player's speed and route accuracy (for receivers), and even break down a player's performance against a specific opponent.

Traditional sports training has been limited by data latency and scarcity. If data on a tennis player's swing efficiency and speed were even available, it was anything but real-time, reducing its value as a training tool. Now, real-time data—for example, statistics such as launch angle and escape velocity in baseball—has the potential to change coaching and athlete evaluation by professionals and fans. You can train and teach players based on precise data about their speed, explosiveness, arm strength, accuracy, and more. You can engage the audience in myriad ways. For instance, it is only a matter of time before this becomes a central part of real-time fantasy sports.

This kind of tracking could also be deployed in a wide range of public facilities, such as airports. Passengers could agree to be tracked so they don't miss their flights. Tractors or deicing trucks could be tracked on the tarmac, both for inventory reasons and to prevent accidents. With plentiful data, needs could be anticipated and met before they became problems.

The Smart Factory

That's the consumer world, but how might the Immersion of Things play out in the practical, results-oriented world of business? Because engineers and designers are still just beginning to explore the potential of the previous stage, the Intelligence of Things, we lack any concrete examples of companies who are leveraging this pervasive, immersive, responsive computing environment. However, we can offer some educated, highly realistic speculation on what will very likely be possible in the very near future.

Join us, if you will, in the factory of 2030.

The manager of this factory runs a highly automated operation. All the machines are highly instrumented and connected to a virtualized version of the factory. For every real machine on the factory floor, there is a virtual avatar of that machine in the immersive computing ecosystem (likely a private or public cloud). The state of each real machine is recreated in its virtual doppelganger. For a computer numerically controlled milling machine, for example, its avatar reflects every possible physical state that can affect its operation, including:

- The position of its actuators
- The current spindle speed
- The tools in the tool carousel
- The temperatures of the bearings
- The temperature and viscosity of the oil
- The level of the coolant
- The state of the machining operation
- The computer numerically controlled instruction being run
- The fixtures and jigs being used
- The state of the bed in terms of metal chips from the cutting
- The acoustic emission of the motors driving the stages
- The current in the motors
- The state of the part (with a dynamic computer-aided design representation of the ongoing cutting operation—sort of like an animation)

In fact, the virtual factory in the cloud recreates the real factory as completely as possible. This includes "digital twins" (a term of art for mediation, along with "cloud things") of all the machines; the positions and states of the helper robots; the positions and states of the automated guided vehicles (AGV) that ferry the tools, fixtures, and jigs in the tool crib; the flow of parts through the factory; the inventory in the incoming buffer; the inventory of finished parts in the outgoing buffer; the environmental state of various parts of the factory including air quality and temperature (important in precision manufacturing); automatic part inspection using a variety of means, including probes, optical, industrial MRI, and manual; and the state of the infrastructure of the factory, including moisture in the flooring, automatic fault circuit interrupters in the electricals, and heating, ventilation, and air conditioning (HVAC) health, which ensures the consistent temperature and humidity levels necessary for precision manufacturing.

This extended virtual factory also includes the supply chain by connecting, in a mediated way, with equipment such as third-party shipping vehicles. The system achieves this by communicating representations of each shipper's state across the cloud. This data includes the contents and location of a truck bringing supplies for the next job and the status of the truck scheduled to pick up supplies from the output stage (an overfilled or late truck could block production).

A combination of real-time, analytical, and machine learning algorithms continuously examine the virtual factory and pull in other data: the lead times needed to deliver shipments, commodity and utility pricing, and the weather (a snow storm may hold up a shipment). With this data, these algorithms fulfill a broad range of management duties. They control factory operations, including giving AGV commands, matching AGV and robot instructions to the part being manufactured, and making dynamic adjustments to overcome unexpected events (machine breakdown, AGV failure), and replanning after same. They assess the health of machines, setting and managing the schedule for changing tools, replenishing or changing consumables such as oil and coolant, predicting failure, and maximizing machine uptime.

General Electric, one of the largest and most influential companies on the planet, is already moving to offer some of these solutions on an industrial scale. GE's far-flung footprint, its efforts to think of its machines as services rather than as products, and its expertise in industrial control make it an ideal company to target varied industries with IoT. GE calls this "industrial Internet of Things." However, GE has taken its strategy a step further, offering up industrial-level digital twins in the form of Predix, its cloud-based "platform as a service" offering that enables industrial-scale analytics for companies seeking asset performance management.

Predix has become a centerpiece of GE's digital strategy. The information Predix can gather, the service level agreements it will likely guarantee, and the predictive capabilities it will offer in the future all describe inversion on a massive scale. GE has further strengthened its position with a series of acquisitions including ServiceMax and BitStew.

Predictive Maintenance

Machines today are purchased and used while giving little or no feedback on their condition to those maintaining and relying on them. Problems are fixed either through routine preventative maintenance or through reactive problem solving. A broken machine leads to costly, disruptive downtime as users wait for repair and can lead to other cascading failures. This is bad for the machine, its manufacturer, and the user.

What is needed is not the machine itself, but the outcome: a pump delivering a liquid to a location where it's needed. In other words, reliable industrial performance as a service. This is inverted thinking as we describe it and as practiced by many forward-thinking businesses: reimagining not just their products but the entire nature of what they deliver and how they provide value. Predictive maintenance, which forecasts likely component failures within a practical time window, is essential to this reliability.

That observation led Saar Yoskovitz to start Augury Systems, which equips the pumps, chillers, and fans in a facility's HVAC system with external sensors, with the heavy data processing work done in the cloud.

Predictive Maintenance (continued)

This allows Augury to do predictive, condition-based maintenance, which is different from preventive maintenance (which simply flags parts that need to be replaced based on their effective life) or reactive maintenance (which flags failures after they occur). The company works with Trane, Carrier, Johnson Controls, the Brooklyn Navy Yard, Aramark, universities, data centers, and others.

"If it's a million-dollar piece of equipment, why not buy it as a service?" says Yoskovitz. "Augury is indicative of a movement: Object as a Service, or OaaS. Why would you ever buy something like a dishwasher again? Why not just get dishwashing as a service?" In such a model, the consumer would not own the dishwasher; instead, the appliance, soap, regular maintenance, and monitoring via IoT would be provided by a third-party company for a monthly fee that might vary based on the number of loads the consumer washes. That's an example of the inversion model detailed in chapter 2: greater flexibility, more actionable data, and needs met in ways that give the customer control and choice.

In addition, these algorithms assess the load balancing of the factory to maximize machine utilization, minimize inventory, and minimize risk in case of failure. They optimize operation schedule and part routing to increase profit and reduce risk. They play out scenarios involving critical parts (e.g., "If machine 2 breaks, your most profitable part will be blocked"). They infer work-in-progress and apply management's preferred control policy, whether Kanban or constant inventory work-in-progress. They apply statistical process control and learn the correlations between causes and faults.

The algorithms schedule operations to reduce utility costs (e.g., electricity slab pricing, when to use renewable sources or local generation) and schedule maintenance (machine 2 won't be used on Friday; have it serviced then). Their fine-grained level of control increases sustainability (no heat treatment until Wednesday, so we'll turn the furnace off automatically, but preheat it in advance) and safety (e.g., air quality measurements, leak detection from pressure vessels). The system even addresses supply chain issues (e.g., automatic ordering of coolants, safe disposal of effluents, short supply of needed stock material for product 1 due to snow storm, so switch to product 2 setup).

Factory and warehouse applications such as these are every bit as important and transformative as are consumer applications. After all, factories and warehouses are staffed by people, and people will seek delightful, immersive work environments just as they do home environments.

The Human Role

With this system at his fingertips, the factory manager notices a quality problem in one of his runs: the surface finish has deteriorated. Looking at history and at shared learning across multiple factories, the system produces a set of possible diagnoses:

- The coolant could be low or contaminated.
- The coolant level sensor may be stuck.
- The cutting surface of the tool may be worn.
- The stock material may be harder that it was supposed to be.

The manager waits until that the next part is undergoing the same cutting operation, grabs an infrared camera, and dons a pair of *Terminator*-style smart glasses (for example, Microsoft HoloLens). When he sees that there is no coolant level problem, he points the infrared camera at the cut. The system recruits the image, knowing where he is positioned on the factory floor and where he is pointing the camera, and informs him that the surface of the cut is hotter than it should be. With his HoloLens, he is able to see a multilayered view constructed from sensor data and computer-aided design drawings to look inside the machine—an immersive experience akin to "x-ray vision." When the tool is back in the carousel, he gestures to activate an external camera, which pans and zooms and shows him that the titanium aluminum nitride tool has lost its characteristic gold color, indicating premature wear. This is the source of the defect.

He makes a voice note of the problem, which is automatically sent to his tool supplier. With a gesture, he then indicates that the backup tool should be used. As he walks back to his office, a floor moisture sensor that has been complaining for some time uses geolocation to note that he is nearby and reminds him to have a look. He notices a drip from the HVAC system overhead and makes a note to have it repaired. The virtual factory system asks him through his headset if a vacuum/mopping bot should clean up periodically so no one slips on the damp spot. He agrees verbally and gestures to indicate the area that needs to be cleaned. He scans the room to see if any other sensors, also geolocated, call for his attention. None do, and he heads back to his office.

Beyond the Factory Floor

An extraordinary scenario . . . but not out of reach. Every type of technology that makes our imaginary factory function in a seemingly autonomous, responsive manner is under development and active testing today in corporate and academic settings around the world. The implications are profound. We have been talking about inversion as a business model in which engineers and designers begin their work not by thinking about how to make new products to put in the pipeline but using IoT to meet customer needs in new ways, including needs that customers might not even know they have. That has gotten us to the Intelligence of Things. Now we are talking about technology that does not simply meet needs but *anticipates* them.

The potential of the systems presented in the factory scenario extends far beyond the manufacturing world. The same technologies could be used as follows:

- An aircraft maintenance engineer could use augmented reality glasses to see the internal state and history of a turbine and its components. He could see, for example, the ages and maintenance histories of internal parts and know their real-time states through sensors inside. For example, the DAQRI Smart Helmet is a combination safety helmet and augmented reality headset that workers in oil and gas, automation, and manufacturing can use to see virtual instructions, safety information, training materials, and visual mapping over specific reality data.
- A nurse could monitor the condition of patients and the states of equipment, instruments, and medications in the hospital rooms, operating

theaters, and intensive care unit from her station. For example, the Advantech Intelligent Nurse Station System uses a computer dashboard that enables patient monitoring, records management, medication management, ward management, and staff management.

- A personal care worker could monitor the condition and activities of elderly patients either at home or in elder care facilities using a battery of equipment including wearables, pill trackers embedded in "smart prescription bottles" that inform them when (and if) patients take their meds, and cognitive measurement systems.
- A retail employee could monitor stock-outs on the shelves, inventory in the backrooms, electronic price labels, receiving information, reordering information, product expiration, and theft. For example, a "smart store" solution being developed by SAS lets retailers analyze shopping center and store traffic in individual locations or across multiple locations. In one application, retailers can use video or Wi-Fi foot-traffic monitoring to see whether customers spend substantial time in a product area and either direct an associate to help that customer or use that data to modify a store layout to increase sales.
- A building manager could monitor everything from the structure's integrity, the HVAC system, and the elevators to the fire detectors and the basement parking for safety, performance, and cost. For example, using the Clockworks building analytics software we mentioned previously, KGS Buildings was able to help MIT monitor seven thousand pieces of equipment in its HVAC systems, analyze fourteen million points of data per day, and ultimately save $1 million per year through more efficient HVAC systems management.[4]
- A city official could manage everything from the municipal water system to public transport, street lights to traffic flow, from the lighting systems in the city's parks to the security systems in the schools from a single location. Several cities around the world are piloting programs to use sensor networks to provide real-time measurements of carbon dioxide, nitric oxide, particulate matter, temperature, pressure, and humidity. These data allow city managers to take action to reduce pollutants by using such measures as restricting drive times or prohibiting wood fires.

Predictive Health Monitoring

There's a lot of talk about wearables and much excitement about devices such as the Fitbit and the Apple Watch. But wearables can be much more than luxury items, especially in the treatment of medical conditions. The adoption of wearables will take parallel paths: adopted by consumers for general health, fitness, and activity monitoring, and adopted by consumers with medical conditions to improve their lives and perhaps even save them.

Take epilepsy, a frightening condition. To be able to use sensors to detect, react to, and, most extraordinarily, to prevent seizures is a powerful example of the potential of the Intelligence of Things, which is indicative of the solutions developed by IoT health company Empatica, which makes wearable devices that forecast neurological events.

Built on analytics running in the cloud, the Empatica system transmits data from the wearable, through a user's smartphone, to the Internet. The solution is designed to detect episodes of chronic neurological conditions such as epilepsy by measuring electrodermal activity combined with movement data and using real-time machine learning to send an alert to a smartphone and predict the onset of a seizure. As shown in 1996 research by Mangina and Beuzeron-Mangina, electrodermal activity—the skin's conductance of electricity—correlates to activity in several regions deep in the brain, including the amygdala and hippocampus. Research published in 2015 by Dlouhy et al.[5] showed that a seizure that strongly stimulates the amygdala can cause the patient to stop breathing. Monitoring is a way not only of predicting adverse events related to such conditions as depression, schizophrenia, and dementia, but also of saving lives.

"If somebody goes to check on a person during or after they have had a grand mal seizure, then they are less likely to die," says Rosalind Picard, chief scientist. "In some cases, simply saying the person's name or turning them over (gentle stimulation) might save their life. Anybody could do this potentially life-saving action, they just need to know to go check on the person—don't leave them alone right after a seizure."[6]

This same approach might also herald other predictive tech for health care. Why can't similar systems be used to predict the next migraine or attack of clinical depression? That is the future of the Intelligence of Things: anticipating acute episodes and even preventing the worst effects of chronic neurological diseases.

- A farmer could use *precision agriculture* to monitor the moisture, pests, and mineral content in different parts of her field, dispensing irrigation, fertilizers, and pesticides in as precise, timely, and economic manner as possible, particularly tuned to both microclimate and weather. For example, venerable manufacturer John Deere is reinventing itself as a smart agriculture business by offering such services to farmers to whom it previously only sold tractors and combines.

As you can see, there is no industry, no area of business, and virtually no arena of human activity that will remain unchanged by the introduction of these technologies—not just as devices but as responsive, personalized, immersive *environments*.

Building the Immersion of Things

In many ways, this has already started happening. Early applications have started appearing. For example, automaker Hyundai recently added a skill to the Amazon Echo that allows users to start, lock, or unlock the car or turn the climate controls on remotely using Hyundai's Blue Link system.[7] A few years ago, this was a baked-in function that involved a preinstalled, hard connection to a remote control. Now, by recruiting Alexa's functionality, Hyundai has made the feature wireless and voice-controlled—a faster, more intuitive solution.

Then there's the deal between DHL and Smart to use the Smart car's "smart" trunk as a secure delivery point, allowing customers to either pick up parcels or drop off parcels for pickup using their smartphones.[8] These types of solutions will connect and proliferate as a patchwork. Over time, Intelligence of Things thinking will make the components of cars, appliances, machine tools, hospital equipment, HVAC systems, and forklifts more open, more accessible, and more recruitable, enabling designers to compose scenarios that transform productivity, safety, and sustainability.

Meanwhile, the burgeoning maker culture is breeding a cadre of designers, programmers, and hackers who are doing interesting new things in a massively parallel way. But do they work for you? Do you have the right people?

Most importantly, immersion requires a focus on the total customer *experience*. In the Internet of Things, communication with everyday devices allowed a business to have a direct connection to the user, and in the Intelligence of Things, the everyday devices could be recruited to new tasks. In the Immersion of Things, the consumer stops being aware of the boundaries of each device and instead is immersed in a total experience. To build this experience, it is not enough simply to recruit the devices that surround the user. *Every* available resource must be recruited, and this includes all of the non-device data that a business may have. In an immersive shopping experience, the business must not only orchestrate the various devices involved—the user's smartphone, RFID tags on nearby products, digital merchandise displays in the store aisles, and similar items—but also the consumer's shopping history, wish list, payment methods, and so on. All of the architectural components we've described—discovery, context, orchestration, machine learning, collaboration—must be harnessed to achieve the total experience.

Safeguarding IoT Devices

The Immersion of Things brings with it even more novel challenges to our ideas about security and privacy. We have already mentioned the distributed denial of service attack on Dyn, which hosts the Domain Name System. The distributed denial-of-service attack brought down major websites such as Twitter and Netflix and revealed a harsh reality of the connected world: we live in a world filled with insecure devices. Devices we barely think of as being connected, from baby monitors to toys, have the potential to be turned into digital attack vectors if they receive the right lines of code from a malicious third party.

That was one example. Medical device maker Animas, a Johnson & Johnson subsidiary, provided another. In 2016, the company alerted patients using its OneTouch Ping insulin pumps that hackers could potentially exploit a lack of encryption in the communication between insulin pump and the remote blood sugar monitor and direct the device to deliver unauthorized doses of insulin—a dangerous flaw.[9] Johnson & Johnson warned customers but stated that it knew of no examples of attempted hacks on the device. For something more recent, consider the news from February 2017 in which Germany's Federal Network Agency (in charge of the country's telecommunications oversight) banned the Cayla talking doll because of concerns that hackers could steal personal data and eavesdrop on children by recording private conversations carried out over the doll's unencrypted Bluetooth connection.[10]

For events with more serious implications, we bring you the 2015 hack that struck multiple power distribution centers and brought down several electrical grids in the Ukraine.[11] Or Stuxnet, the computer worm first identified in 2010 that was used to sabotage centrifuges at an Iranian nuclear facility.[12] The implications are clear. The next war may be fought without a single bullet fired. Remember, IoT has not appeared suddenly out of nowhere. Instead, it has crept up on us because for years there have been industrial systems controlled by computers—systems that we think are isolated but are not. The cyber-physical world of today involves big things such as power plants, dams, and power grids; critical things such as door locks; and dangerous things such as cars and furnaces.

This is not a call to panic or retreat from innovation. Far from it. It is a call, however, to be aware of the risks and realities. More to the point, to adhere to some basic wisdom regarding the connected, immersive things that are becoming ubiquitous:

1. *Prevent unintended consequences*. Sanjay is a DIY home automation aficionado. When he purchased an Amazon Echo, he connected it to his home's door lock and developed a habit of letting people into his house by looking out the window and shouting, "Alexa, unlock door!" That is, until he realized that on a summer day, a thief could easily shout "Alexa, unlock door!" through a window and gain access to the home.[13] That's but one example of the many unintended consequences that could (and undoubtedly will) appear as IoT matures and propagates. The Immersion of Things world could yield a plethora of unintended hazards, such as accidentally starting your car with a verbal command while it is in the garage, leading to a dangerous buildup of carbon monoxide. Such consequences need to be assessed methodically by mapping out all the possible device states and making sure states that users deem undesirable cannot be

achieved easily. This is called *verification*, a field of research in logic synthesis for semiconductor chips and a rich area for development in the IoT world.

2. *Watch for "Trojan horses."* Every day, we purchase devices containing hardware, firmware, and software that came from uncertain provenance. Some come from companies and suppliers that can be trusted; inevitably, some come from companies and suppliers with negligent or nefarious intentions. In the IoT age, we cannot continue to be cavalier about the sources of our hardware, software, and connectivity. Very soon, we will seek to establish secure sourcing in the same way we work to ensure that counterfeit prescription medications are kept out of the U.S. supply chain. As the Dyn attack and other attacks have made clear, our devices can be turned against us to effects ranging from the annoying to the destructive. Our primary defense is vigilance.
3. *Cognitive firewalls.* A standard security measure in computer networks is the firewall—a system that monitors traffic according to preset rules. A firewall is important in the world of Immersion of Things, but it may not be enough. Consider the following example. Your wife calls you and leaves you a voice mail that says, "Running an errand, left food in microwave for the kids, please turn it on for three minutes." You recognize your wife's voice and follow her instructions.

   Now let's say your wife calls you and leaves the same message—except that she inadvertently says to turn the microwave on for 300 minutes. You have a sense of the physical properties of foods and microwaves. You question this command and refuse to execute it. But what about a "smart microwave"? Would it simply heat the food for five hours and as a result, set the kitchen on fire? With current technology, it's quite likely. Existing systems have a model of a microwave but not a model of the real world, so they have no context with which to analyze the validity of such a command. As a result, they execute it literally.

   A cognitive firewall (a concept pioneered by Josh Siegel[14]) is a watchdog of sorts that protects us from such misunderstandings of instruction and context. A system that applies and simulates a physical model of the real world, learns from the past, and applies a watchful eye over the physical world and the consequences of commands, will save the cyber-physical world from both intentional and unintentional damage. If a command does not make sense based on its model of the real world, the system will conscientiously object to it. Building this level of machine intelligence will involve advanced control theory such as observers, estimators, machine learning, statistics, and more, but it is under development even as you read this.

The Great Shredder in the Sky

As we have discussed, privacy is about policies and behaviors more than technology. For example, in an immersive world where the practical boundaries between devices no longer exist, who owns the data? This is a huge question.

The way privacy works right now, you can't remove your data from social networks, for example. You can't pull it off Facebook and put it into a new social media platform like you're extracting a PDF file because Facebook owns the data. That's the business model of a lot of companies: make their sites sticky and get your data so they can own you.

If businesses are to give consumers choice and control in the new, connected world, they can't do it halfway. Walled gardens of data simply won't work anymore when devices and systems interoperate. If you own a car that talks to the cloud, and you want to move to a different cloud server, you should be able to extract your data and then move it to another server, just as you would a cartridge. This "data cartridge" idea is speculative, but it is coming. People will demand the right to control their data—or even destroy it. Imagine a cloud shredder, the Great Shredder in the Sky.

There will be serious pushback over this from businesses whose business models depend on controlling part of their customers' data. However, given sufficient demand, those objections will fail. Remember that mobile phone carriers pushed back against number portability for years, insisting that if they could not keep their customers captive by forcing them to change cell numbers if they changed providers, their businesses would wither and die. Well, number portability became a reality with the Telecommunications Act of 1996 and the major mobile carriers are still doing just fine.

One of the other privacy questions looming over IoT now relates to the ability to "hide" from immersive technology. In the RFID world, we developed opt-out capacity in RFID tags; companies or even customers could use "kill" commands to keep objects with RFID tags from being read or tracked. In this new immersive world, perhaps there needs to be a "hide" command that would allow any individual who does not want to be immersed in the fog of responsive sensors, actuators, and computing to opt out—to hide from detection by the coffeemaker or thermostat. If anonymity is a basic human right, and that right is threatened by the IoT, then this is a huge issue we should be discussing *now*.

The Revolution Will Be Human

As you can see, as the technology-driven environment around us becomes more intuitive and responsive—more human-like—there will be a correlating increase in complexity and ambiguity. There will be threats and concerns—security, physical implications, unintended consequences—but there will also be untold benefits. For example, IoT is a powerful asset for the disabled. Prosthetics researchers have made great progress in connecting the electric signals in muscles to prosthetic limbs. Recently there has been brain control of orthotics as well.

However, if the Immersion of Things is to have practical applicability, it must be as easy and frictionless to use as in the film *Minority Report*—minus the unnerving Orwellian implications. Interfaces will become more important than ever, more fluid and natural and immersive, developments that users will welcome so long as we have verification systems and checks and balances in place.

The revolution will start with buttons. It will then include speech, thanks to cloud-based natural language processing systems. Bluetooth headsets, perhaps custom-built with longer battery lives, will make it easy and inexpensive to stay in contact with our constellations of devices at all times. Displays, especially wearable displays such as the

HoloLens, using technologies such as liquid crystal on silicon, will change how we see the world.

Each of these developments represents an opportunity for businesses to evolve from meeting needs to anticipating needs to curating experiences and creating new markets in areas where none previously existed. The businesses that lead the way in the Immersion of Things will need to adapt to a world where privacy and control over data is an absolute right and companies are rewarded not for locking their customers into product ecosystems or walled data gardens but for delivering choice and meeting needs.

Immersion and Inversion

In this chapter, we have painted a portrait of a world that does not yet exist, but is coming soon. It is a world in which, in many instances, the product-first thinking that characterized the pre-IoT era will not only become obsolete but irrelevant. In an environment where connected devices and immersive intelligence learns user patterns and preferences and orders products preemptively, before they are needed, the drive to purchase and consume will evolve.

Certainly, for everyday needs, consumers will shift their attention from acquiring objects and services to obtaining and designing experiences. Institutional customers, from banks to universities, will turn their focus to immersive environments that deliver cost savings, sustainability, customer satisfaction, and efficiency. A fully inverted economy is one in which meeting needs—both anticipated and unanticipated—is the definition of value.

We have presented our case for the engineering foundation of inversion and the evolving nature of IoT that will power it, illustrating the technologies that will be necessary to realize both the Intelligence of Things and the Immersion of Things. In the final section of this book, we pivot to the business implications of this new economic reality. We will explore how businesses can leverage the technology of the IoT, Intelligence of Things, and Immersion of Things to successfully take their place in the inverted, connected economy.

III Inversion and Business

7 Inversion in Practice

Reinventing Your Market and Your Business

The evolutionary process from the Internet of Things to the Intelligence of Things to the Immersion of Things is about more than technology. It is about the three-stage evolution of business that's happening right now: from selling products to meeting needs and, finally, to enabling experiences. But the process of inversion—of transformation from a company working in a self-limiting product-first environment to one that plays in a wide-open landscape of experience-based areas—requires doing more than investing resources in IoT and thinking about new business models. It requires that your company *become* something new.

To see what we mean, let's talk about Sears.

Started as Sears, Roebuck & Co. in 1886, the company once dominated the American retail landscape. By the 1960s, it was the largest retailer in the world, and by the early 1990s had annual revenues of nearly $60 billion. Sears had launched one of the first commercial Internet plays, Prodigy, owned Allstate insurance and the Dean Witter Reynolds stock brokerage, and had a brand that was trusted by millions of American consumers.[1] However, by 2015, Sears's sales had fallen to $25 billion—a 50 percent decline from 2006. By the beginning of 2017, the company had lost more than half its value, and as of October 2016 its total debt was more than seven times its market cap of approximately $600 million.

Meanwhile, from 2006 to 2015, sales at upstart online retailer Amazon.com rose from $10.7 billion to $107 billion.

Why didn't Sears become what Amazon has become? How did it take such competitive advantages as a nationwide customer base, ample cash reserves, and a deeply trusted brand and squander them to the point of irreversible decline? The long answer is better suited for a business school graduate course, but our concise answer is that as the world changed around it, Sears was unable (or unwilling) to adapt because it did not divorce itself from the business model that had sustained it since the Grover Cleveland administration. In inversion terms, Sears was unable to change from a *traditional* company to an *inverted* company, leaving it vulnerable to agile, needs-oriented, innovative competitors such as Amazon and Wal-Mart. Clearly, success or failure at inversion is not a function of size or lengthy corporate history. Something else is at work.

To this point, you have read about business models, architectural principles, and the transformation of the IoT into an engine for creating immersive experiences. Now, as we transition to the final section of this book, four key questions must be addressed:

1. What do we mean by traditional and inverted companies?
2. What are the implications of inversion and IoT for business?
3. How can a business and its leadership put together inversion and IoT to assemble a clear picture of the path forward?
4. What are the essential steps in moving from traditional to inverted?

In answering those questions, let us first assert that the fate of Sears is a cautionary tale, not destiny. However, in the years to come, it will become a matter not only of relevance but of survival for businesses to adopt new models and open themselves to disruptive new visions. One of those models is inversion. As for visions, read on.

Traditional and Inverted

In this context, a traditional company is a product-first company that has achieved consistent growth, predictable profitability, and substantial market capitalization by making and selling its products using a time-tested model of design, fabricate, package, market, ship, sell, service, and repeat. It is likely to be a large, established company, but startups can also operate with a traditional model. Traditional companies generally do one thing exceptionally well—developing and selling office productivity software, for example—and for a long time, that was enough. However, that is no longer true.

Inverted companies are disruptors and renegades. They are typically lean, comfortable with risk, and are often young companies or start-ups. However, they don't have to be start-ups, because this category has nothing to do with scale. In theory, there is nothing stopping large legacy corporations from operating as inverted companies. What inverted businesses have in common is that they adhere to the inversion model and have a corporate vision based on meeting needs, not on pushing any one particular kind of product. In the connected economy that is rapidly taking shape, inverted companies are positioned to capture more and more of the market share previously presumed to be inescapably in the hands of large traditional corporations—exactly as Amazon did with Sears.

When we're not writing books, we spend a great deal of our time working with traditional companies making the transition to inversion and with innovative inverted companies. Start-ups and newer companies run by youthful entrepreneurs, such as those popping up in the sharing economy, naturally tend to be inverted. Disrupting existing business models is part of their DNA and how they compete, and their comfort with IoT, the cloud, mobile apps, and the other technologies that enable inversion make an inverted business model a natural, even inevitable, choice. This chapter's primary concern is how large, established traditional businesses facing potential obsolescence can become inverted businesses by adopting inversion and leveraging the power of IoT.

Anecdotal evidence alone—the rapid rise of agile IoT start-ups, the profusion of IoT devices and the businesses built around them—supports the idea that this transition is a business imperative. But data make an even stronger case. In 2014, a PriceWaterhouseCoopers's report on the sharing economy (a textbook example of needs-over-product inversion) found that revenue in just five important sharing sectors—automotive, hospitality, finance, staffing, and media streaming—was expected to grow from $14 billion to $335 billion in 2025. More to the point, the same report found that while revenues from traditional car rentals were expected to grow only 2 percent from 2013 to 2025, revenues from car sharing were expected to increase by 23 percent. Traditional DVD rental revenues? Down 5 percent. Streaming entertainment? Up 17 percent.[2]

Those sharing economy increases will not occur in a vacuum. They will occur because agile, tech-smart companies following the inversion model take market share from Hertz and Redbox. Extend the same dynamic over every sector of the marketplace, from heavy industry to aviation to retail, and you have a picture of what the next twenty years are likely to bring for traditional businesses that are unable to adapt to the connected world.

However, we are extremely bullish about the ability of traditional companies to adapt. Traditional companies already have vital advantages over their smaller, leaner IoT competitors. They're already players in the industry where they generate most of their revenue, so they have market share and customer relationships. In addition, they have scale. They have resources and experienced personnel. They have a supply chain and a sales channel. They may have access to capital. The inability of many traditional companies to make the transition to inverted companies has little to do with material causes.

Why do so many seem unable to escape the gravity well of corporate inflexibility, irrelevance, lassitude, and ultimately, bankruptcy? We believe it's because of three basic obstacles:

1. An inability to envision their business as being anything other than what it is today—an example of frame inertia.
2. Confusion and fear over transitioning to an inverted, needs-first paradigm.
3. Resistance to change.

Collectively, the authors of this book have worked in the worlds of business and technology for decades and remain active players in technology and innovation. We helped to create the RFID industry and IoT. We have seen a panoply of strong, established businesses miss opportunities because they made the deliberate choice not to consider the idea that their traditional advantages would enable them to dominate as needs-based inverted companies.

Think for a moment about Nest and the typical utility company. Utilities have a direct link to our homes; why didn't one of them ever offer a Nest-like device to connect people's homes? Who knows more about a home's energy usage than the utility that delivers that energy? As we see it, it is long past time that the people reading meters were connected—that meters and grids themselves were smart and connected. However, the utilities that have tried to implement a "smart metering" strategy have mostly fumbled the ball. For example, Pacific Gas & Electric, which distributes power to northern and central California, promised its customers that they would benefit from lower costs when it ran a 2009 pilot program testing connected usage meters. But the opposite happened: usage numbers and bills went up for many customers, and utilities in other states proceeded with "smart grid" programs with excessive caution.[3] Only within the last few years have utilities succeeded in installing smart meters in substantial numbers—about sixty-four million by 2015.[4]

Nest did this. Then Google bought Nest Labs, and now they're the leader in smart, connected utilities. How did this happen? An inverted way of thinking dominated a traditional industry that had every possible advantage.

Take property sharing. What type of large business has a massive database of information about homes and their condition, size, and security features? Insurers, already some of the largest businesses on the planet. Why didn't Travelers, Liberty Mutual, Allianz, or some other insurance giant leverage its data and its trusting relationship with property owners to create what Airbnb did? Because, just like hotels, they chose not to think beyond their current business model due to the fear of eroding their existing business.

But inversion is not limited to start-ups. If you operate a large traditional business, you can and should become an important player, a driver of innovation and growth, and even a disruptor in many vertical markets in the connected world of IoT. It's likely that the only thing stopping you is your mindset.

What are the implications of inversion and IoT for business? Survival and prosperity versus lost opportunity—at best. You can be Amazon,

Airbnb, and Netflix. Or you can be Kodak, Research in Motion, and Nokia. Let's look at how to achieve the former. Marc Gorlin got a call from a tile setter working at his Perdido Key, Florida, condominium. Some of the tiles for his bathroom had broken in shipping, and it would take several days and an additional shipping charge to get new tiles from Birmingham, Alabama. Gorlin, a serial entrepreneur who cofounded online lender Kabbage, figured there was probably someone driving out of Birmingham at that moment—headed for Perdido Key—who wouldn't mind hauling a box of tile and making a few bucks. It was the lightbulb moment that birthed the company Roadie.

Gorlin launched Roadie to capitalize on the vast, untapped cargo capacity already on the road. The company started with $10 million in venture investment from sources such as TomorrowVentures, the UPS Strategic Enterprise Fund, and the Mellon Group.[5] Initially focusing on college students—who are mobile, eager for extra income, and comfortable with technology—the company has begun to disrupt the $90 billion U.S. shipping industry.

With more than 25,000 registered drivers, Roadie runs on the strength of its smartphone app. Drivers post their destinations, customers post "gigs"—a destination, the things to be delivered, and their size—and the system acts as matchmaker. Senders can specify delivery times or choose "whenever," and they don't have to box up their stuff; fees are based on the size of what's shipped. Gigs are automatically insured, customers can track progress in real time from the app, and drivers ("Roadies") get on-the-road discounts and support. Roadie is turning a previously existing resource into a marketplace based on Gorlin's vision.

Inversion Roadmap

1. *Identify product-first business lines you could convert to needs-first.* Figure out areas of customer need that you can address before considering products that you can sell.

2. *Reimagine your company as a player in the inverted world.* Adopt a "launch and learn" approach that invites the failures that lead to innovation.

3. *Open source your talent.* Recruit talented people rather than acquiring them.

4. *Embrace the "hacker mentality."* Be open to reengineering, reverse engineering, and exposing the insides of products to make them better.

5. *Determine opportunities that can "layer" over existing products through IoT.* Look at ways you can add non-connected products to the connected world using solutions such as RFID.

6. *Think like a start-up while leveraging traditional company assets.* Cultivate creativity, disruption, and quick action.

7. *Launch and learn, test and evolve.* Iterate quickly and learn from failures.

8. *Innovate, don't acquire.* Free your people to create new solutions. You can't invert through acquisition.

9. *Make inverted world bets: early and often.* Take risks and don't worry about what doesn't work, only about what you learn.

10. *Change how your customers perceive you.* Make sure to communicate your new direction to your customers; their long-standing perceptions of you probably won't change on their own.

11. *Ask "How do we play in the inverted, experience economy?"* Instead of asking about what you can make and sell, focus on creating experiences.

An Inversion Roadmap

What must an organization do to successfully make that transition from traditional to inverted? Before we dig into recommended steps, it is important to state something that may be obvious to some but not to others:

Virtually every legacy business that transitions from traditional to inverted will spend considerable time operating as both.

In other words:

The change from traditional to inverted is an evolution, not a radical discontinuation of existing business lines and customers.

Something must generate revenue and pay the bills that allow companies to venture outside of their historic core competencies. For example, venerable glass manufacturer Corning has turned heads with "smart glass," a technology made for IoT that transforms a variety of surfaces into functional touchscreens. However, at the same time, the company still makes optical fiber and cable for the communications industry among many other products—because they continue to be highly profitable. Transitioning to inverted does not mean burning down your existing business, just changing focus.

For example, Corning made the Gorilla Glass that Apple used for the displays in its earlier generation iPhones. Because of this, Corning had an incentive to begin thinking in an inverted way—addressing Apple's need for lighter, more durable displays—rather than simply pushing product—but traditional thinking prevailed. Corning stopped innovating and delivering additional value and Apple decided to switch to sapphire screens from GT Advanced. However, it's possible to recover from an inversion misstep. When Corning introduced Gorilla Glass 5, thinner and more scratch-resistant than its previous materials, it won back Apple's business.

Let's figure out what makes an inverted company tick and walk through the step-by-step process that a traditional organization should follow to productively begin the transition to inverted, with the understanding that it is possible to operate for years, even decades, as both. As long as you have a strong presence in the inverted, connected world so that you remain relevant in the needs-first economy, there's nothing wrong with being both.

Step 1: Identify Product-First Business Lines You Could Convert to Needs-First

Often companies spend most of their time looking at new iterations of the same product, which effectively prevents them from seeing beyond the business model needed to produce that product. To be fair, that approach is why major companies have been so successful in developing so many products of surpassing quality and reliability, including appliances that can last for decades without a major failure.

However, for computer manufacturers, software manufacturers, operating system manufacturers, and database manufacturers, while that approach kept them as important players in their space, it prevented them from coloring outside the lines.

That's the first step. Any company looking to make the transition to inversion must think about where their future lies—even if that future looks unnervingly different from the present. What could you be doing? What needs could you be meeting and turning into revenue streams by tapping into the technology of IoT? Forget the impossible; everything is impossible until someone does it. What do you want your business to be in twenty years?

Step 2: Reimagine Your Company as a Player in the Inverted World

In the process of achieving the impossible, there will be many point failures—or what we like to call *points of success*. Even a chain of individual failures, if experienced in the fearless pursuit of innovation, often leads to discoveries and new ideas that end up producing successes. When those occur, you must be able to use them as opportunities to color outside the lines. That's not only desirable, but risk and failure actually increase a company's credibility as long as it quickly pivots to other equally daring offerings.

One example is Amazon's Fire tablet, an iPad competitor that debuted in 2014 to great fanfare, and then failed miserably. However, when it became clear that the Fire was a flop, Amazon didn't retreat; that's not the Jeff Bezos modus operandi. Instead, Amazon immediately rolled out Echo/Alexa, which has not only been a market-changing sensation but has created a whole new customer vocabulary and in-home experience.

This was not an accident. It happened because Amazon and CEO Bezos understand that failure and innovation go hand in hand. By jumping back into the market with a new needs-first solution, they reinforced their image as a bold incubator for daring, creative ideas. If they had viewed the Fire's failure as reason to stay away from innovating for a few years and lick their wounds, someone else might have beaten them to Echo. Amazon's culture and confidence in their ability to bounce back from a single failure freed them to make a big bet on Fire and then, when it didn't work out, pivot immediately to the next idea. How could you breed a culture that doesn't fear or punish failure but uses it as a springboard to the next big thing?

Step 3: Open Source Your Talent

As you begin to color outside the lines, it is possible that you won't have all the in-house talent to get where you want to be. This is an opportunity to recruit talent into the company, instead of acquiring a company that may clash with your business or just saying, "I guess we're not going to play in that area."

In the inverted world, that talent can come from many sources and many geographies, and could simply be widely distributed contractors completing tasks. The point is to leverage the best talent for the need at hand in an agile, flexible, open, and collaborative way, not necessarily to build your roster of permanent employees.

Step 4: Embrace the "Hacker Mentality"

In the inverted world, the hacker mentality is normal and acceptable. What is the hacker mentality? It's the attitude that things were meant to be reverse engineered, taken apart, and put back together in new and better forms. It's not only agile but works toward results through speed, not perfection. This mentality allows inverted companies to introduce new products rapidly and evolve at a fast pace.

Step 5: Determine Opportunities That Can "Layer" over Existing Products through IoT

This usually causes companies a lot of concern. They're departing from their core business, which is what they've spent decades perfecting and where they have made a lot of money. This process leverages existing, profitable business lines and their customers in order to "layer" related IoT plays over them—a strategy that can yield incredible growth. Apple's entire rise to becoming the world's most valuable corporation came as a result of this process. Apple was a computer company that decided to disrupt music. Then they became a phone company. Then they created a market out of thin air for this thing called a tablet. Now they are reinventing wearables. All because they decided to venture into areas that were outside of their core competency—sometimes far outside it, and filled with ferocious entrenched players—and see if they could disrupt those markets by finding bold new ways to meet customer needs.

Complacency due to the existence of a stable revenue stream built on a product-first model is a trap for many businesses. In other words:

In the inverted, connected economy, the fastest way to kill revenue is to cling to it.

If you remain exclusively a traditional player, you might potentially make a lot of money for a long time, but you will eventually be replaced by an inverted company offering the customer more choice, more control, and a more beneficial experience. Remaining traditional is a choice, but an unnecessary one. Virtually any business can move to an inverted model.

Step 6: Think like a Start-Up while Leveraging Traditional Company Assets

We read a great deal about start-ups and their disruptive, world-changing ideas. But traditional companies have all the resources they need to become inverted—and to do it in a much bigger way. Why don't they? Why is Amazon the leading player in drones? Why aren't airlines innovating and planning to become major players in drone delivery of freight, cars, or even passengers? Why are they not building the first unmanned aerial vehicle flight companies?

If Delta is unable to think of itself as more than just an airline, it limits itself to coloring within the lines. But if it starts looking at *airspace* as its business—not just airplanes—many possibilities open up. An airline brings tremendous knowledge of the Federal Aviation Administration, maneuvering, safety, aerospace engineering, and beyond. If part of the future is a world where there will be multiple layers of unmanned aerial vehicles, drones, and manned aircraft moving through the same airspace, who better to own it than an airline? Why do we think that Elon Musk is the only one who can think about this sort of thing?

In another example, companies like Terrafugia and AeroMobil, not to mention Airbus's Vahana initiative, recently announced prototypes of flying cars.[6] Why didn't GM or Volkswagen get there first? In February 2017, a Chinese start-up called eHang announced that it would be testing the first-ever passenger drone service in Dubai.[7] Why didn't United Airlines get there first?

Think about Hyperloop and the rail industry. Hyperloop is the brainchild of Musk's SpaceX, a transportation system that would propel a pod-like vehicle through a tube on a cushion of air at five hundred or more miles per hour. While the jury is still out on the concept, why was it conceived by somebody who manufactures cars? U.S. Rail came up with the Acela rail line that runs between Boston, New York, and Washington, D.C.—that was their version of inversion. Then Musk came along and showed them what "what if" thinking really looks like.

That's the difference between being a traditional and an inverted company. The point is, a successful traditional company can leverage its expertise and know-how to not only compete as an inverted company, but also to lead the field. You don't have to start from scratch or be intimidated by the so-called agile start-ups. But you do have to begin thinking of yourselves as a player in the inverted world.

Step 7: Launch and Learn, Test and Evolve

The roadmap for success is to test new products, launch them into the market, learn from failures, iterate, and evolve rapidly. This is often done without the user even being aware that products and services are evolving. For example, Lyft recently introduced a new feature to its ride-sharing app: Round Up and Donate. Passengers can choose to round their fares up to the nearest dollar and donate the extra cash to charity. Lyft customers don't see the feature in development; it simply appears on the app one day at the end of a ride, and users have to activate it before they can use it.

Solutions that evolve as a result of software updates represent unique challenges for inverted companies. They must make efforts to communicate new features' existence and value to customers, often through multiple channels, in order to have customers appreciate the innovation. A box that arrives in the mail can't be ignored; a seamless software update is easy to miss.

Step 8: Innovate, Don't Acquire

You can't invert through acquisitions. We are not categorically opposed to acquiring other companies, but it should be done for the right reasons. An acquisition can give you access to useful technology and new talent, but it will not—and should not be expected to—transform your company's culture or how you think about your mission.

Instead of trying to buy inversion, dust off your file of "high-risk" projects. Identify three to five new projects that are completely different from your current business. This means being willing to spend money—not treating these ideas as pet projects, but investing for the long term to figure out how to make them work as products. Develop a new architecture. Build prototypes, watch them fail, and then build more.

Too many large tech companies view research as a source of intellectual property, put it in a lockbox, and throw away the key, taking it out later only to be sold as an IP portfolio. They don't apply it, which robs the world of some incredible advances. For example, while we were writing this book, physicists at IBM Research developed a previously impossible triangular molecule that could be the key to quantum computing, an experimental type of computing that, if feasible, would make use of quantum mechanical states such as superposition to perform calculations at far greater speed than even today's fastest transistor-based computers.[8] Imagine if discoveries like that were rapidly tested, prototyped, and productized.

Many traditional companies tout the billions they spend on research and development, but that is misleading. A great deal of that money is spent on incremental improvements of products they already have, which is necessary to generate the ongoing revenue that supports the often-risky bets that inversion demands. It is often necessary to maintain traditional business while testing the waters as an inverted business, doing both in parallel. However, the ongoing traditional revenue stream should not be allowed to breed complacency toward the eventual need for the entire business to transition to an inverted model. That guarantees failure. Before you know it, you're in full retreat and someone else is leveraging the opportunity that should have been yours.

Step 9: Make Inverted World Bets: Early and Often

Leaders of traditional companies often stay away from making bets. They use terms such as "shareholder value" to insulate themselves from the consequences, so no one expects them to even venture into new areas—including themselves. Elon Musk announces Hyperloop, and everyone says, "This is fantastic, how will it work?" But if Amtrak held a press conference and said, "We are going to build a system to move passengers from Boston to Seattle in six hours," no one would know how to react, because that doesn't happen. There's nothing stopping traditional companies from making big bets—they have billions in available cash, resources, and customer bases that small tech companies can't hope to match—so why don't they?

Struggling to recover from its emissions scandal, Volkswagen is preparing to make a dramatic bet. In 2017, the German automaker will launch Moia, a fifty-person IoT ride hailing start-up meant to go against services such as BMW's ReachNow, a Zipcar analog that allows customers to borrow BMW vehicles on an as-needed basis. The independent spinoff's near-term goals: develop an affordable ride-hailing service and create an IoT carpool service, something it calls "connected commuting."[9] That is a bet.

Not making bold moves becomes a kind of groupthink, a self-fulfilling prophecy. Corporate leaders worry that they will appear desperate, harm their credibility, crash their share price, or simply fail. The simple equation is *no bets=no wins*.

Step 10: Change How Your Customers Perceive You

In the traditional business model, you grew a customer base, got those customers to think of you as one thing, and kept those customers forever. In the inverted world, you're changing how customers perceive you and what you are capable of. As you transition from traditional to inverted, you will challenge preconceived notions about your business and what it can be.

Once you determine where you will play in the inversion economy—and where you envision going tomorrow—take control of how customers view your business. Communicate with them about your plans. Reassure them that you will not suddenly discontinue the products they have come to depend on. Tout your new ideas and your evolution into a collaborative partner that will meet their needs and give them more control over their experience.

These are functional shifts in mindset that must happen in order to shift from traditional to inverted, and not all companies can manage the cognitive leap. Why do established hardware companies fail to start cloud storage businesses that wind up surpassing them in valuation? Why do established shipping companies, with scope, customers, and cash, fail to see the untapped local delivery potential in personal vehicles and launch their own sharing economy delivery plays? It is difficult to escape frame inertia; doing so not only demands that you think of your company in a different way, but that you help your customers do the same.

Step 11: Ask "How Do We Play in the Inverted, Experience Economy?"

The traditional question has always been, "How do we build that?" Now, it's possible to do it without having any product other than the architecture that ties other system together. You can orchestrate without ever releasing a new physical product.

Ride hailing companies didn't design the GPS system. They don't design cars. They doesn't employ drivers. They orchestrate and architect an environment that brings other things together to get people or parcels from point A to point B. Airbnb takes assets and people and brings them together. Inverted companies must be masterful orchestrators.

Playing in an area of the marketplace becomes a powerful question in the connected, inverted world. Let's deconstruct the question, "How do we play in that?"

- *How*: What strategy do we use (not what do we manufacture)?
- *Play*: Participate, be active, hopefully become a leader.
- *In that*: The future of that sector of the connected world.

With this question, the other questions change. "What people do we bring in? What assets do we already have that we can bring to the table? How do we find other players that add to what we're trying to do?"

Now, you're no longer surrendering to the start-up. Instead of developing proprietary systems, you are assembling smart systems. Winning in the inverted world requires agile orchestration over building another expensive product. In other words:

> *Playing in a space is about going from building to assembling, owning to orchestrating.*

Multidimensional Evolution

When you finally turn your people and resources loose, hold on tight. When you become a needs-based company, you will evolve rapidly and multidimensionally. Needs are constantly changing and provoking you to change to meet them. The net result is to become an experience-based business. Consumer product companies, banks, insurance companies, service companies—they are all seeking to understand and evolve based on what the customer expects to experience. Customer expectations drive everything.

It is no longer an option to remain exclusively a traditional company, pushing products into the pipeline. Today, companies that understand and can shape and curate the customer experience dominate the economy. For example, cosmetics companies are multi-billion dollar conglomerates that have traditionally operated by developing products, running massive branding campaigns, and selling through retail. Now the same companies are struggling because they do not understand the real-time needs of their customers. In the past, the way to deal with that was to flood the market with thousands of beauty and grooming products.

But now we have L'Oréal, once a traditional company manufacturing conditioners, shampoos, and hair color. Now the company is putting "smart brushes" in the hands of its customers, using a variety of sensors and cloud computing to analyze the texture and moisture level of the user's hair condition and recommend specific products to improve its condition. That is a traditional company in an extremely traditional market with the vision to reimagine and reinvent itself on the fly, even as it remains a traditional company. That's inversion. (We will dig deeper into L'Oreal's vision in an interview in chapter 8.)

Chief Inversion Architect

Companies of any size intending to make this transition from traditional to inverted need to establish a new position: chief inversion architect. (For obvious reasons, we're going to pass on using the position's acronym.) Companies have had chief technology officers for decades, and chief innovation officers since the mid-2000s, and both are important in the world of IoT. The CTO focuses on determining what technologies will best serve the company's mission and acquiring them, while the CINO scrutinizes trends, determines which align with that same mission, and helps move people and resources toward developing new solutions in those areas.

However, neither the CTO or CINO are charged with questioning an organization's *mission* itself. That is precisely the role of the chief inversion architect. This job is about giving a company *permission* to invert. This job is to manage the transition of traditional companies from a product-first model to an inversion model and to overcome the outside-in perspective that suggests that innovation and disruptive vision can be acquired. Microsoft, which has gone the inversion route, still fell prey to this mentality when it bought Skype and LinkedIn, acquisitions that have continued to operate as satellites but had little to do with Microsoft's transformation into a more needs-oriented, experience-oriented business.

The role of the chief inversion architect is to champion and shepherd true inversion from within because they see it as a necessity for survival. First, this person should be able to see far beyond the limits of the company's current business and be comfortable with disrupting that business—even going so far as to create what's known as a Culture of Disruption.[10] What are the frontiers of the company's current business and how are they changing? Does the company's current conception of itself and its mission still make sense, or is it limiting the company's ability to meet unmet needs? For example, Tesla may have started as a company trying to disrupt the auto industry by building a better electric car, but it's clear that Elon Musk now sees Tesla as a power company. That change in self-concept has opened vast new markets in which Tesla can be a player.

The chief inversion architect's second role is to determine the next steps the organization should be taking to move toward that inverted business model. That type of near-term/long-term strategic thinking means this individual needs to have a strong grasp on a wide range of disciplines, from Internet and IoT to software development, economics, organizational management, and product development. The chief inversion architect should also possess a clear understanding of the company's culture, its potential to limit innovation, and how it can be transformed. We will discuss this more in chapter 9.

Third, the chief inversion architect is the point person for determining the resources, people, and planning needed to turn an inversion strategy into a revenue-generating line of business. What needs to be added to the company and what departments, product lines, or initiatives should be discontinued? Finally, the chief inversion architect is responsible for developing the operational, financial, and technology roadmap that guides the company's evolution and creates new business opportunities.

This does not imply that every company must hire a new C-level executive to serve as their inversion guru. We appreciate that many businesses will balk at creating a new technology-oriented department and a new C-level position in addition to the CTO, CIO, chief security officer, and CINO positons, all of whom may well be engaged in inversion-like strategy. Instead, because inversion is first and foremost about a vision for your company, it is quite possible for a CTO, CINO, or other skilled individual, who can look past the current mission to see what the company can be, to fill this role. Whether a chief inversion architect is a new position or a single person is unimportant; what matters is the vision to imagine what is possible in a needs-first economy.

X Marks the Spot

In the previous three chapters, we explained the evolution of things from connected devices to interconnected devices to immersive experiences. Now, it falls not only to the chief inversion architect, but also to every other person who shares in the company's vision—from CEO and CTO to CINO, engineers and designers—to answer this question: Where does your business sit on that continuum *today*? As you explore the potential risks and rewards of playing as an inverted business, where do your technology, expertise, creativity and market allow you to compete now?

Are you solidly within the Internet of Things, with devices that can be equipped with sensors and actuators and turned into sources of useful data? Do you possess the software engineering chops to work in the Intelligence of Things, using applications to recruit hardware to create value in new areas? Or could you be positioned to stride boldly into the Immersion of Things with technology that powers seamless, intuitive experiences?

Where do resources, imagination, and opportunity collide today?

You Are a Visionary

The steps represent a valid process, but in all honesty, they are not enough. For corporate leaders who aspire to innovate, their greatest obstacle tends to be themselves. *You have to know how to get out of your own way*. We tend to have rigid expectations of what we are capable of doing. We impose limits on our vision and creativity and deify those who don't appear to have those limits.

We all have the capacity to be visionaries. The main quality that sets visionaries apart from the intelligent CEO running a successful traditional company is the nerve to look at where to play, accept the risk, and make big bets. Only self-imposed limitations keep the rest of us from doing that. This book attempts to unleash the power of traditional companies on the inverted, connected world. When that happens, the rate of innovation will skyrocket.

Consider Google. They were a software company, a search company, a mapping company—but not a driverless car company. Until they were. First, they rolled out a few vehicles in 2009 to take photographs on college campuses. Then they started testing them on streets. Fast forward a few years and we have Waymo, Google's stand-alone self-driving car play. Now, the software and search company is a force for innovation in a field it wasn't even in a few years ago. Could Honda or Ford do that? There is no reason they could not.

Inversion Is About How You Define Yourself

Inversion is a way of thinking. It is about how you define your company. If you define yourself according to a product—a storage company, a rail company—you have inherently limited your potential, where you can play. Instead, it's time to start training yourself to think differently. Inversion works for small, medium, and large companies; legacies; and start-ups. How can you invert what you do, meet needs, and innovate? Stick your toe in the water, and if something fails, it fails. You go back, iterate, and continue placing multiple bets. This is not a game won with a single bet. You do not transition from a traditional Sears to an inverted Amazon with one initiative, no matter how brilliant it might be. You get there one step at a time.

After all, maybe Hyperloop is not the answer in transport. Maybe it's drones. Perhaps multiple bets will tell us that airspace is superior to rail and that building the structure for Hyperloop is just too costly. Maybe there's another answer no one has thought of yet—but it requires that you begin to ask, "What if?"

With decades of experience working across many verticals around the world, we believe without doubt that traditional companies can not only play in the world of inversion, but dominate it. You have the experience and the know-how, the access to resources and talent. All that's missing is the thought process. Amazon isn't afraid of failure. Why should you be?

We believe that all traditional players are equipped to become inverted. We've assembled a few that are showing the way; you can read about them in the next chapter. These changes will occur as a result of mindful interplay between large companies and start-ups, making this the most exciting moment in the history of technology. But it will only happen when traditional companies begin their evolution to inverted companies.

So partner mindfully. Become a player. Create the connected world.

8 Conversations with Inversion Players

In this chapter, we share interviews conducted with a selection of large corporations that are leading the way toward inversion as powered by IoT. Some are consumer-facing companies that are making the change to an inverted model, explicitly reconfiguring areas of their businesses to operate on a needs-first basis. Others are leveraging IoT to better meet their customers' current needs. A third group of companies is developing the essential infrastructure that will enable other businesses to more effectively practice inversion.

Some additional notes about the interviews:

- They are the product of original live interviews and conference calls.
- They have been lightly edited for clarity.
- We have deliberately chosen a diverse cross-section of *Fortune* 100 companies.
- We have not been compensated for any of these interviews and we disclose any involvement with the principals.
- We have deliberately kept the conversation broad and allowed our interviewees to share their current and long-term visions with us.

Schlumberger

Interview with: Neil Eklund, formerly chief data scientist, Schlumberger. Interview conducted by telephone.

Summary: Schlumberger is focusing its IoT efforts on "asset health management"—the remote monitoring, diagnostics, and prognostics for a company's physical assets. This work exemplifies remote sensing as a classic way of exploiting the IoT in an industrial environment to meet customer needs in ways that offer greater utility and value. However, the intent to make an eventual transition to the Intelligence of Things and a consumer focus is clearly reflected in their thinking.

Q: *What is asset health management?*

A: You've got assets in the field and if they fail, it causes a problem. With aircraft, it causes delays and cancellations. In the oil field, it causes delays in drilling. So you reduce downtime and the amount of equipment you need. You also save on secondary damage. The idea is you can spot failures several days out. If I tell you a part is going to fail in five minutes, that's too late. You should have replaced it a week ago. If I tell you it's going to fail in six months, it's way too early.

Q: *What kind of data are you gathering on something as complex as an oil rig?*

A: You've got three types of data. You've got environmental data so you know the temperature of the water and the air, and the temperature on the ground. You've got control systems data. And you have usage data—the history of the asset itself. Sometimes, there aren't a lot of sensors on a system for asset health management. Things like ground-based gas lines might have them, because weight is not an issue. Aircraft engines don't have them because weight is a huge issue. The IoT effort in industry is so new that we're not putting sensors on things that are specifically for asset health management. Instead we are just using some of the data that's available. You don't get insight into every possible failure because of that.

Q: *Now I have the data, and I'm going to use it to predict in real-time, versus having designed systems a priori with the notion of doing that. That has deficiencies, because I might not be collecting at the right frequency or be able to look at the right things. How much of asset health management now is accidental IoT versus intentional IoT?*

A: It's entirely accidental. You check inside a pump and you happen to collect the control systems data for years, but don't really know what you're going to do with it. Then if you have a failure, you say, "Let's go back and see what was the cause," instead of trying to predict the failure. That's a huge issue. It happens all the time, on all sorts of platforms. It happens at GE, and Pratt and Whitney, at Honeywell—it's everywhere.

Q: *What is your vision of where highly distributed asset management science could be heading as we get into a hyper-connected world?*

A: Machines stocked with sensors that can look for specific faults and do it much, much better. Schlumberger is already putting PHM [prognostics and health management] in the design process to make machines that are smart enough to have sensors that will be good for predicting faults. I've spoken to some guys from companies doing even more maintenance, and they've got sensors that talk between cars and compare notes and look at diagnosing faults that are specific to local driving conditions. It's remarkable stuff.

I think the future for IoT and industry is just more deliberate sensoring of assets. Flight restriction or data restriction, certainly that's an issue, but many, many assets—cars and trucks, and gas lines and all sorts of things—can operate just fine with another pound and a half strapped to them.

Take driverless cars, which are going to be a very big deal. Undiagnosed failure in those will be a big issue. You need the ability to tell at a distance if this asset is operating properly. I've seem efforts in software PHM: you monitor your software to see if it is operating properly. For driverless cars and other autonomous vehicles, this will be a big deal. This year at the PHM conference, we're doing a track on pilotless vehicles that will be critical for the future.

Q: *What is your vision for the future and how do you expect machine learning to change to accommodate it?*

A: The model today is plainly centralized monitoring. That's why asset health management started out in aviation: it's expensive. But as PHM moves into cheaper and cheaper objects, it's in a car that might cost $40,000 today, but someday it will be in your refrigerator. We need to push the monitoring software from a "human in the loop" centralized model to a distributed model where each car, refrigerator, and asset of any sort has asset health management software built right into it. When nothing is amiss, it pings the mothership, "All is well." If something is wrong, it sends back a ping that there's a problem. One central monitoring system is not going to work for many assets. It will have to be pushed out to individual devices.

Today in asset health management, there are two models. There's the hybrid approach, which works pretty well. You've got old school engineers who want to build a physics-based model that's very precise but very specific to assets. If you change anything, you've got to change your whole system. On the other hand, you've got the machine learning people who say, "Give me some data and I can do anything." I am the latter.

How will machine learning change? In the [19]80s and early [19]90s, machine learning was based on taking many, many passes over the data. You'd build a neural network and look at the data from those passes and tune your weights. There was your model. Then in the 2000s, people came up with a lot of algorithms that were sort of a single pass. You wanted to look at a stream of data going by and try to infer something from that. That drove innovation in modeling. The interesting thing about IoT is if you find one fault in one asset, you can immediately look for that signature in all of your other assets. With a decentralized approach, if one asset learns a particular fault, then you can figure out a way to distribute that knowledge across all of your assets.

Q: *We are living in a highly connected universe of highly disconnected interests, without standards that enforce what those connections should be. What is one thing that reflects your optimism around this future connected world, and one that reflects your concern?*

A: The concern is privacy. IoT is creepy in a lot of ways. Your phone is tracked and it's with you at all times. Your insurance company can take the data from your car and say, "We bought this data on a secondary market, and it says you're a dangerous driver so we have to bump your rate." How all of this data is used and who has control of that data—those are the kind of things that concern me.

On the positive side, reliability. With airplanes, delays and cancellations will be reduced, and cars will be more reliable. That's the dream of asset health management. Unquestionably, we are making strides toward that.

Insights: Schlumberger's current efforts are typical of industrial applications of IoT. The technical focus is on remote sensing, while the business focus is on reducing costs by enabling faster, more targeted responses to problems in the field as well as anticipating and correcting failures before they occur. Despite the company's vision of applying such sensing technology to consumer products such as refrigerators, Schlumberger is unlikely to become a consumer appliance manufacturer.

However, if consumer appliances become Intelligence of Things devices (as we predict in chapter 5), Schlumberger will be able to recruit appliances into its remote monitoring services simply by downloading suitable software—a clear example of recruitability allowing an inverted company to extend its core competence into markets that would have been hitherto impossible to penetrate.

L'Oreal

Interview with: Guive Balooch, global vice president of L'Oreal Technology Incubator. Interview conducted by telephone.

Summary: L'Oreal is bringing consumer beauty products into the connected world with the introduction of the Kérastase Hair Coach, the first "smart hairbrush." Cosmetics and other beauty products may be physical objects, but L'Oreal is using inverted thinking to reimagine its mission, taking it beyond simply launching product and instead delivering valuable personalized data that enables customers to purchase beauty products with precision.

Q: *What is your vision for L'Oreal and where it fits into IoT?*

A: My job is to think about how connected devices and new experiences will transform beauty. I think that in the field of beauty, consumers who are interacting with beauty products and buying beauty products are rapidly changing because of the technology and connected data all around them. The future is really about how we can create new, personalized, customized experiences using things such as IoT and machine learning and neural networks—to figure out how we can give consumers the best products for their needs. This is such a product-driven industry that I think people are going to use the connected world to get better products and products that are more personalized through innovation—not only in IoT, but also in the supply chain.

Q: *Could you talk a little bit more about performance and IoT in the beauty world, given the extensive portfolio and diversity of products that you have?*

A: Think about how women today get the perfect foundation for their skin. Fifty percent of women cannot get the perfect foundation because there are thousands of potential skin tones visible to the human eye, but only a few colors available at the point of sale.

When you think about consumer expectations, at the end of the day what women really want is a product that matches their skin as perfectly as possible. If we could use measurement technologies and data to understand more about the true skin tone of customers, we could start to personalize products for them and get [the products] to them faster. We're launching thousands of products in the beauty industry and there are only so many we can launch to meet every consumer expectation. Now, expectations are rising because we're able to get so many customized experiences in every field.

Q: *What is L'Oreal doing in IoT?*

A: On my team, we're trying to create new connected objects, taking devices that are typically used for beauty routines, such as brushes, and figuring out how we can connect them for the beauty consumer and bring them value around coaching. We're using sensors in devices that are typical in a beauty routine and figuring out how we can use data to coach consumers on getting the best products, whether it be on the hair or skin.

We are also using wearables such as our UV [ultraviolet] patch to communicate and educate consumers through IoT around the level of exposure they're getting to environmental conditions. We want them to be educated about their daily exposure to pollution or UV and to figure out how to use the right products and when to use those products during the day to keep their beauty routines perfect. We're also finding ways to connect all of these different objects and devices and to understand better through machine learning how to develop better products through customization, new supply chain innovations, and the ability to get better understanding about consumers.

Q: *The brush, which you introduced at the Consumer Electronics Show, is unique. Can you describe it?*

A: More than 50 percent of Google searches for beauty are related to hair. Also, people say, "I had a bad hair day." They don't say, "I had a bad skin day." Hair is really important to people's lives, and the environment and different attributes of exposure can affect how their hair looks and feels. Our idea was to partner up with a start-up that we felt truly understood how to bring consumer-centricity and beautiful industrial design to create a connected object that would not add complexity to the consumer's life but bring them information about their hair.

The brush itself has been designed to be very beautiful, on par with $200 brushes on the luxury brush market. Then we partnered with Withings to create the hardware. Our professional hair brand, Kérastase, helped us do the marketing and branding and it's their product.

The brush itself has six sensors in it. It has a microphone, so as you're brushing your hair the microphone hears the acoustics of the hair and uses our algorithm to give you information on the condition of your hair—damaged, dry, affected by conditions of the environment. It has a force sensor in it, which tells you how tangled your hair is, which today is very important. When people have wavy hair and they come out of the shower, they need to use the right products after to ensure that their curl is intact and their hair is not frizzy.

It has an accelerometer and gyroscope to understand the location, and we have a thermal conductivity sensor that measures the moisture in the hair. It has Bluetooth and Wi-Fi, so you don't need to have Internet [access] at the moment to use it. The goal is, by using this hair brush every day, it gives you information about the quality of your hair, and then you can get the right products. We tested the brush with polymers on the surface of the hair and have been able to detect a dose-dependent effect. We're able to see if a leave-in conditioner is no longer on the hair, because as a leave-in conditioner goes off the hair, the acoustics as you're brushing will change. You will have more friction between the hair fibers as the conditioner is wearing off the hair, and the conductivity in the sensors that helps you measure that.

Today, a lot of women don't know what kind of product to use. They don't know if their hair is unruly, or really [how] frizzy their hair is, or how tangled their hair really is. Six weeks after using the brush, you can go back to your hair stylist and have some information to share with them about how your hair has been evolving the past few months. Then they can give you the right product and the right way of styling your hair. It's really just a coaching tool using a beautiful device that people normally would use at home.

Q: *The potential of IoT is to understand the consumer far better, be able to sell more to the consumer, and please the consumer. Yet the rate of introduction of technologies has been very conservative. What should be done to change that in the beauty industry?*

A: In the medical and fitness industries, there's a direct correlation between information and product efficacy. It's not only a direct correlation, but also something that's very one-to-one in terms of interaction between the consumer and the product. The beauty industry is more emotional. You can have a product that's efficacious, but you won't like it because it's not sensually correct, doesn't smell the way you want it to or doesn't look the way you want it to. Because of that, it's been difficult in the past to figure out how to bring consumers added value through things like IoT.

Now, I think we're seeing a larger inspiration taken from the connected health industry, and also the fact that consumers will use technology for every product in their lives. Because of that, we're seeing more situations where we're trying to figure out how to connect the beauty world. My team has a really unique view of industrial design, data, data science, and machine learning. We also have people within the team who understand beauty consumers, who are very complex. Our industry has been typically very product-driven, more so than many other industries, and the consumer has been conventionally, classically driven by the sensory and emotional aspects of products.

Because of that, it has taken a little bit longer to adapt both the consumer as well as our industry to the changing world when it comes to this type of information and technology. We definitely now see, with all the innovations around, its potential. There's huge potential for using IoT to give consumers the best types of products for their needs and educate them so they don't use the wrong product ever again.

Insights: L'Oreal recognizes that beauty is a product-driven industry, but nevertheless applies inverted thinking in seeing the potential for technology and IoT and connected devices to transform consumer experiences. They apply IoT technologies directly to familiar products such as hairbrushes, but critically, connecting such products is only the means to an end.

The ultimate goal is to use IoT to give consumers more data that they can use to make more informed choices and buy the best possible products—something that is especially important with products such as cosmetics that cannot be customized digitally. This is a holistic application of our key points from chapter 2—using sensor-gathered data to give customers more choice, control, and an enhanced experience—that points the way to the same path for thousands of consumer products companies.

SoftBank

Interview with: Amit Kumar Banbey, chief scientist at SoftBank Robotics. Interview conducted by telephone.

Summary: The ne plus ultra of the Immersion of Things is the prospect of socially intelligent, humanoid robots, and SoftBank Robotics is advancing robotics to create robots that can perform physical tasks and interact in a human-centric environment.

Q: *In this connected world, we're going to have virtually everything talking to everything. So what is that connected home going to look like, or feel like with respect to the role that humanoid robots play?*

A: From my perspective, you have a connected ecosystem, which is mostly static without robots, and robots are going to be serving as the body language of this connected ecosystem. They will be the physical capabilities that elevate the capabilities of this connected system. There is a smart home and the robot is part of that ecosystem, where a room has sensed that there is some problem with the human and robot is the first responder to go there and solve that problem with the person. This could take place in the home, offices, public places. The intelligence-connected sensors can say that there is some problem, and the robots can go there and address it. There is enormous potential when robotics and IoT come together and function as the brain and the body.

Q: *We're entering a world where technology is going to be a big part of everyday life. Could you give us a definition from your point of view of a humanoid robot?*

A: It is not really well-defined and concrete, but what I see is that first, a robot has to have some autonomy and motion. That is very basic for me. Then, if you would look at the robot, you should be able to match some of your physical appearance and features with the robot. That is, for me, the humanoid part. If it has some way of expressing with body language, then it can be considered as humanoid. A humanoid robot has some sort of human-like way of being connected.

Q: *What are the three biggest challenges that have to be solved in order to make socially intelligent robots work?*

A: There are basically three sectors: connectivity, sharing of information, and collective intelligence. You have information in there but you need to share in a way that makes sense out of it. That is the place where there is a real need to develop things. The information sharing has to follow privacy and other critical concerns, and that is very challenging as well. So, once this type of ecosystem is there, the challenge is going to be how to make (robots) not learn bad things and behave in a better way.

Q: *So humanoid robots might be fulfilling needs that humans can't do for themselves, such as maintenance in tight spaces in the home or assisting an elderly or disabled person who has a fall or other emergency that the intelligent home detects. How far can intelligence in robotics go?*

A: There are limitations in the sense that the computation on any robot is limited, unless it is connected to some cloud or IoT system where the computation can be elevated. Computation is the technical requirement to have the higher level of intelligence, because you have to process lots of data and make decisions. A stand-alone robot can have a limited intelligence but enough intelligence to do and make some decisions in a local situation. We should be able to identify that there is a person in need [and] the emotional state of the person, and [we should be able to] support them.

That is local decision making. For this type of decision making, the robot won't stand around without connecting with the patient. But if there are some bigger problems, the robot needs to see things which he cannot see personally from his perspective. Then we need the higher level of intelligence, and that's where developing intelligence comes in. Decision making is from collective observation and then everything is processed collectively.

Q: *You talk a great deal about social intelligence in robots. What are the current limitations on achieving this kind of intelligence?*

A. The problem is with how the robot is able to perceive the environment and make sense out of that. With the whole computer or camera, what they see at a basic level is something safe. And then the robot has to figure out, "Okay, this pixel is basically a chair, or a board, or something." It has to make sense of what it sees at the object level and then it has to go to an even higher level of persistence and make sense of the situation based on the things it is seeing. This type of multilayered persistence is still very limited in terms of technical advancement. Even if we can achieve a perception level of two-year-old child, that will be very good.

Insights: The Immersion of Things is the most ambitious stage of IoT evolution, and robots that can interact meaningfully with humans is one of the most immersive experiences we can imagine. Robots fill an important void in the immersive experience: despite a profusion of sensors, actuators, and computing power, robots will be the only components of the immersive environment that can perform tasks on an ad hoc basis, tasks unanticipated by programmers. As such, they will be invaluable assistants in creating the type of industrial experience described in chapter 6. SoftBank is diligently pursuing that level of socially intelligent behavior.

Johnson & Johnson Supply Chain

Interview with: Mike Rose, vice president of Supply Chain Visibility. Interview conducted by telephone.

Summary: As the world's most comprehensive health care company, Johnson & Johnson's Supply Chain (JJSC) is leveraging IoT to obtain the data needed to benefit customers, leading to a revolutionary concept: the "Internet of Health."

Q: *Can you talk to us about your work at JJSC to automate the supply chain using IoT?*

A: We're tracking and tracing our products in the supply chain for a number of reasons. One is to provide better security around the supply chain so that patients receive genuine product[s]. This is not a huge issue in the U.S. or in Europe in particular, but there is a higher prevalence of counterfeit drugs in the emerging markets. Track and trace is becoming a price of entry in almost every market now as consumers and regulators are becoming more aware of the concerns around counterfeit products. The other area we're working on is upstream traceability, again driven by patients and customers. They want to know the source of a product. Is it genuine? Were the raw materials in that product sustainably sourced? An example would be in the case of our product line where we use palm oil. They want to know that we're working with farmers who are harvesting palm oil from the farm itself and that we are following sustainable practices.

What does that mean from an IoT perspective? Traceability and visibility are critical for the IoT. For instance, whether it's sugar or palm oil that moves from the farm to an aggregation point to a processing point, [it's important] to have visibility to where it came from and into what product batch it was dispensed. This is now a requirement for many of our raw materials.

What's also important is moving product traceability downstream to the distributors and the pharmacists. Safeguards are needed in the supply chain so that counterfeit products don't have the opportunity to get introduced into the legitimate supply chain. This ensures greater stability and integrity of the supply chain through traceability. Distributors who are buying from us and distributing product to pharmacies or hospitals are doing so from a secure supply chain.

This creates a barrier of entry for the counterfeiter. With a secure supply chain, they will bypass pharmaceutical products and medical devices.

Q: *How do you see IoT changing how patients and customers experience your products and health care products in general?*

A: Over time, IoT will enhance the customer experience with our products. It's going to take some time but I can see a day where more and more people use their devices to interact with intelligent or smart packaging. For example, when patients take prescription drugs, prevailing industry averages show that they are only about 40% compliant with their drug therapy. What if there's a way, with the Internet of Things, of interacting with the patient with intelligent packaging, like a smart blister pack, or partnering with cell phone apps. All of a sudden there's an opportunity, if the patient agrees to opt in, to [help them] become more compliant by being messaged: "Did you take your medicine this morning?"

Johnson & Johnson has a health app called "The Seven Minute Workout." I used it yesterday, so today, roughly around the same time that I worked out yesterday, it'll send me an alert saying, "Hey, time for a Seven Minute Workout." I could see the same thing happening with medication, using an intelligent package with all the proper opt-in and privacy concerns all addressed.

So that's where I think the industry and JJSC are eventually going, and you'll see direct, patient benefit for this. When I look at the health care industry, I think the long-term vision of IoT is that it could help reduce the overall cost of health care. We know that patients who are compliant with their medicines have better outcomes and require less care. That's a win-win from so many angles.

Q: *What's the long game for JJSC and IoT in health care?*

A: Well, expectations are rather high in the health care system. I think you're seeing a lot of experimentation on the part of the Baby Boomers and the elderly, and I think it's only going to increase. The health care system is always trying to reduce cost but deliver the same level of care—if not more, better care. I think technology has a huge role to play there. As you look at Internet of Things, this becomes a very kind of logical extension of that—it could be the Internet of Health.

It's smart sensors, smart devices, it's packaging that has incorporated these technologies, it's also integration of the data, as well as cell phones, the Internet, cloud-based solutions, and electronic medical records systems. With that in place, the natural [solution] is, you've got a cell phone, you've got a package, so you can imagine the various combinations of these technologies coming together. It might start with RFID, but then it will go to smart sensors and then your device—your phone or Apple Watch or Fitbit. As a consumer, you will self-select what combination of technologies and applications works for you.

I'm going to digress into a slightly more futuristic world, beyond IoT, where I have spent the last decade of my career. What happens when we collect all this data and get into machine learning and deep learning? In that world, a doctor knows not only what medication a person is taking but also what he is ordering from the grocery store. It's all done automatically, so my doctor can say, "He is purchasing five pounds of bacon and he is on a statin for cholesterol."

Or suppose that my Fitbit shows metabolic indicators. We start getting into a world where we can start predicting, not just monitoring or analyzing. A doctor can say, "If this pattern keeps up, this is what's going to happen." More important, how can we reverse that? These are coming purely from machine learning and deep learning based on complex understanding of a hugely multi-variant situation. Health care costs should go down as a result of that.

Insights: JJSC is entering the world of connected health care devices via IoT, starting with basic tracking and compliance monitoring. That's important, but its larger vision is the vision of the Internet of Health. It points to a critical reality in IoT. Health care, fitness, the general optimizing of human physical and mental well-being—these are an ecosystem unto themselves with exquisite potential to align not only with the convenience aspects of IoT but the more profound predictive elements of the Intelligence of Things and even the Immersion of Things. All this will come with some profound concerns about privacy and the protection of patient health data, and those will need to be addressed with robust solutions.

Now we transition to conversations with two of the many technology players creating solutions that will power the inverted world. (While it is very possible to pursue an inverted business model without relying on the technology discussed here, we have chosen to maintain our focus on the systems underlying IoT.) There are, of course, thousands of other businesses that fall into this category, but these interviews represent the extraordinary level of innovation occurring throughout the worlds of business and technology.

Velocloud

Interview with: Sanjay Uppal, CEO and cofounder. Interview conducted by telephone.
(Disclosure: Linda sits on the Advisory Board of Velocloud.)
Summary: Velocloud is one of the leading players in cloud-delivered software defined WAN: a cloud network that provides an enterprise-grade connection to cloud and enterprise applications. This is important because the IoT will increase the volume of network-connected devices by several orders of magnitude, which will in turn increase our dependence on reliable connectivity. In a hyper-connected world, you cannot simply rely on one source of connectivity, because if that source goes down, so does your business, hospital, or school. Velocloud is among the companies addressing one of the great challenges that will arise from massive connectedness: the need for redundant, seamless, ultra-reliable, multi-protocol connectivity.

Q: *How did Velocloud get started?*

A: We looked at what was happening to the IT world over the last ten years or so. The cloud has come in, and instead of being deployed on individual businesses premises, IT has moved into the cloud. There are reasons for that. Number one, businesses are more interested in the outcome of the technology that they deploy rather than owning the technology that they deploy. Second, enterprises are much more interested in consuming technology as a service than buying a bunch of products and then hiring a bunch of people to run the technology.

We were thinking, "What about networking? How will all these things connect with one another? Why is it that you need hundreds of people to make sure that the right application is dealt with correctly from a quality perspective, from a security perspective, and from a performance perspective?

What connects your individual clients and users back to those data centers is wide area networking (WAN). Wide area networking is interesting because the networking is inside a building or a campus. This is where all the complexity comes in. You have the Internet and wireless and private networking and all the service providers and even Google and Facebook deploying balloons and satellites. All of that is WAN. From our perspective, wide area networking was very complex. Individual businesses did not want to deal with all that complexity; they wanted to deal with the outcome.

We said, "Let's take the functions of wide area networking that are locked on a business's premises and move those functions into the cloud, and let's run those functions as a service. What was a product and a physical appliance or device now gets converted into a virtual service. That's the essence of what Velocloud is trying to do in the networking space. However, what struck us after a couple of years is that it's not just about users. It's about anything that can be addressed through the network, and in particular the Internet of Things. That could be a sensor, an actuator, or what have you. All of those don't just want to talk to one another, but to the applications that are managing and controlling them. And they need to do that over the WAN.

If we conceive of how these billions of things will talk to one another and to their applications over wide area networks, if the wide area network complexity doesn't reduce by one or two orders of magnitude we will have a complete mess on our hands.

Q: *In this confusing world, where does Velocloud play and what does it solve?*

A: For businesses today, and likely for consumers tomorrow, there are two things we do. One is that all the applications an individual wants to access, whether it's web browsing, ERP [enterprise resource planning, a system that allows businesses to automate many back office functions], or watching a movie, will run on a software-defined network. Once you say, "This is the application I'm using," the network will configure itself using software to recognize which applications an individual user is accessing.

Second, there are several networks you could be using at once. For example, when somebody's on their phone, they either use Wi-Fi or LTE. But there's no technical reason why they cannot use both. Use Wi-Fi when it's good and the LTE when it's good, and when both are good you have a choice. This ability to be able to use many networks that behave differently, but yet satisfy the needs of the user and the application, is the second major innovation Velocloud brings to the table.

In software engineering we have a [saying]: "Any problem in computer science can be solved with another layer of abstraction or indirection." Essentially what we've done is abstracted away from the specifics of each network. All you need to think of now is that they're connecting to your application through the cloud. You don't have to say, "I want to get on my Wi-Fi," or "I want to get on my LTE." You're just connecting to the cloud and everything else is done for you."

Q: *The new connected world will be a highly distributed environment, where you're going to have end points anywhere and everywhere. We're going to have hubs and gateways and ways to bring analytics closer to the edge. Can you paint a picture of this distributed model?*

A: When we say "the cloud is the network," and you look inside that cloud, it's a set of cooperating gateways. A gateway is just a piece of software that runs at various important points in the network. An important point is where a lot of Internet service providers meet. There's a set of gateways that Velocloud deploys around the world in twenty-seven locations, but there are gateways that service providers, known as carriers, telephone companies, and other companies provide. That's the place where these virtual services run.

If a sensor wants to talk to an actuator, some questions are asked. How is the network connectivity to that actuator directly? How good is it if it goes through the closest gateway? Should I dynamically connect from my sensor to that actuator without involving any intermediate layer, or should I go through some gateway because I need a critical service that the gateway is providing? All of this is done in real time on the fly. What is my closest gateway? What services does this gateway have? If it's a service I really need, then I connect to it to get to my actuator. If I don't, then I can connect directly from a sensor to an actuator, from a branch to another branch, or a user to another user.

This gateway concept has taken away all the heavy functionality that needed to reside at the end point—application steering functionality, security functionality, performance mediation functionality—and moved it into the cloud, running it as a service.

This is especially true when we talk about connected devices. If everything were to run through the cloud, there's an issue of latency if you have to control a device a few thousand kilometers away. If you don't need that speed, you could aggregate some of the traffic and make some decisions locally.

This is edge computing or fog computing. Where Velocloud fits in is that we consider that to be another type of network service, provided on top of the cloud as the network. It just becomes a way to figure out what you need. Do you need the service at the edge? Do you need the service from a gateway? Can you connect directly between these edges? Do you need to go all the way back to the application? All of that is done dynamically for you using software.

The software-defined network runs on everything and everything runs on the software-defined network. That really captures what we're trying to do.

Insights: As we highlight in chapter 4, the connected future will feature many different connectivity protocols that must work cooperatively to ensure reliable connections and satisfying experiences for consumers and businesses. Velocloud is adapting to that world by developing connectivity solutions ideally suited to computation at the "edge"—data processing that occurs in computers residing at the periphery of a network rather than in servers at its center. The versatility of Velocloud's approach points the way toward an infrastructure that will power inversion, making networks flexible and reliable, and allowing users to meet many needs using the solution-oriented "as a service" model instead of owning costly hardware and software.

NVIDIA

Interview with: Jim McHugh, vice president and general manager of NVIDIA DGX-1; Andrew Cresci, general manager for NVIDIA's Industrial Sector. Interview conducted by telephone.

Summary: NVIDIA is bringing machine learning and deep learning into wider use in IoT, particularly in industry. Machine learning and deep learning are being adopted by many of the companies we interviewed, and NVIDIA's hardware makes deep learning possible.

Q: *How is DGX-1, which you're calling NVIDIA's first artificial intelligence (AI) supercomputer in a box, impacting IoT?*

Jim McHugh: It's basically using artificial intelligence—or more specifically in the world of machine learning and deep learning—training the applications that bring the intelligence into IoT. It brings that intelligent application to the edge where data is being processed in a lot of IoT scenarios. What we have today when we look at home automation and IoT in general is connected devices that are somewhat smart. They're communicating with each other. They have the ability to do things like track my usage pattern and take action on that.

AI and deep learning have not quite infiltrated that, but when they do, they will radically change it. When you have these devices at the edge, we don't really take advantage of the data. We just say, "Let's take an average sample of the data and take some actions on it." When we have an intelligent application at the edge, it will be learning constantly and taking actions that aren't just scheduling or connecting. We will be doing better predictions, better controls.

Q: *Andrew, shifting to your perspective on industrial IoT machine learning and deep learning, what's your vision of where IoT should be now?*

Andrew Cresci: Where we are today is, you can control something remotely, which is what the Nest thermostat is. What AI is going to do is bring prediction capability, so instead of just controlling an individual device, you'll say, "Well, there are 25 different things that could affect temperature—the weather, the forecast, patterns of behavior" and you want to blend all these things together to make a better prediction for managing temperature.

The example that I use is a power station. It turns out that in running a power station, [utilities] are only using between 2 and 3 percent of the information available to them. Why? They've got ten, twenty, or thirty years of history of how it runs and what are the failure items, so they write an algorithm that says, "Look, it's usually this pump that fails," and can run pretty well with that. But AI gives you the ability to look at all the data. It looks at all the operating parameters around the plant, which might have nothing to do with this particular pump, and it can do a much better job of prediction. Turns out that the pump might have been failing, but it was nothing to do with the pump; it was something to do with the voltage provided to some switch three degrees away from it. They never would have known to find that.

AI lets you pull in all these different data sources and create a pattern that can help you link all these things together. That's the beauty of what AI's going to do. The reason it's needed is that there is just so much data. You cannot do it by hand, and today's technique of doing manual feature engineering just doesn't scale. You need a machine to create these links and connections, and then the humans can step in.

Q: *This seems like a good time to draw distinctions among AI, machine learning, and deep learning.*

AC: Machine learning typically (though not always) is a hand-coded piece of an algorithm. For example, with the thermostat, "Is it over 72 degrees?" is the threshold. Deep learning is quite different. You're not doing much manual stuff at all. You're feeding the neural network all the data. The neural network is doing its own job of creating a picture of the world that enables it to find features. One of the features might be temperature, another might be voltage, another might be wind pressure or something else. It will connect all these dots, so it will understand more of the variables that affect the temperature in the house. It's about connecting many dots that you didn't know existed. Ultimately, that's a computing problem and that's why DGX comes into it.

JM: We're not hand-coding anymore when we leverage deep learning to change the world of machinery. Once you start relying upon the data to develop feature sets and the functionality that you need, we're not going to rely on expert systems where humans are typing in, "If this, then that." It's going to be more about discovery and allowing it to happen organically, because of the AI capability of deep learning.

The AI capability will enable really the machine to have the ability to perceive, to reason. I know some people think that sounds kind of spooky, but I actually think that's exactly what we want machines to do: to have the reasoning that allows them that predictive capability. Once machines can have predictive capability, it's amazing what's going to happen.

Q: *We've had AI around for at least 40 years, and we've seen it go through periods of excitement, then lack of excitement, and now we're in this rebirth phase. Why? Obviously, we have a lot more data and things are more connected. But AI is not something new, and certain elements of deep learning are not something new. Why is today's world more receptive than, say, ten years ago?*

AC: Three things have changed between then and now. Data. There's way, way more data available in the training system than there used to be. Second, the algorithms are considerably more powerful. [Networks themselves have] a memory of what happened in the past in order to be able to better predict the future. But a big one is access to computers, because for the longest time AI has been the domain of academia. It can take months to train a system to be practically useful, and that does not work in industry. Because AI is a math problem. It's statistical algebra. It just happens to run hundreds of times faster on a CPU.

Now you've gone from a training time of four months to four hours. That makes a considerable difference in the way you approach it, and a number of people are going to adopt AI. We're seeing a recursive effect where you can train faster, write new algorithms faster, and apply more data over a longer period of time so it learns better. This is the Big Bang.

JM: I think the arrival of data and computational power allows people to have much deeper networks. When you have deeper networks, you get a deeper understanding. I also think people have gotten smart about what they can use as data. If you're using an engine component, now there's this whole idea of a digital twin, where I can simulate what's going on in the real world using all that data. Neural networks have been trained to actually run test simulations on my digital twin so I don't have to do it in the physical world to make it learn more. I think the marriage of augmented reality and AI is opening up many, many different avenues.

Living in California, how we water everything from our farms to vineyards will become more controlled because we'll water at the appropriate time, not when something doesn't need it. We'll stop watering if there's a puddle forming. That's where the resources of the world can be extended. That's why getting intelligence at the edge and training applications to live there and make decisions is really important. We will save lives.

Insights: As NVIDIA makes clear, the power of deep learning is crucial for extracting the most value out of *connectedness*, one of the inversion principles we describe in chapter 3. The data coming from sensors at the edge is *new data* that companies previously have not consumed, and the quantities can be huge. The potential to know exactly what the customer is experiencing can only be realized if patterns can be discerned in the data; deep learning is the tool that can do it. This is why deep learning is cited in many other interviews, whether for the Immersion of Things or even just to understand remote sensing data at the IoT stage. The fact that deep learning has only become possible in the past few years is due to advances in computing speed developed by companies such as NVIDIA; that is why they are a key player in the inversion ecosystem.

9 The Inversion Triangle

From the inversion model and vocabulary, we have progressed through the stages of technology-driven change—from creating value by selling products, then by meeting needs, and finally by creating experiences. We have explored the steps necessary to move from a purely traditional company to also operating as an inverted company, and we have provided examples of businesses doing exactly that.

The final stage in this journey begins here, with culture. Inversion is not a technology or a strategy, but a business philosophy. For that way of thinking to be resilient and to fuel transformation, it must become part of your culture—of your entire organization. Let's discuss how to make that happen.

Business in the Inverted, Connected World

For most of its modern history, Schneider Electric, a multi-billion dollar French corporation founded in 1836, was focused on selling its customers electrical control systems. But in the age of IoT, its business model has changed. Today, the company is hard at work transitioning its business to a results-oriented model—specifically, helping hospitals, banks, hotels, and other customers practice cost-saving, outage-preventing "predictive maintenance." By equipping transformers, switches, and other components of a building or campus electrical system with sensors that detect variables such as heat, vibration, and current surges, Schneider can help building managers predict component failure before it occurs.

Eventually, the company plans to sell outcomes as its primary business, not components. Instead of selling circuit breakers, transformers, or installation and maintenance services packages as its primary source of revenue, Schneider will simply bill its customers a flat monthly fee per kilowatt hour. If usage rises, the cost rises, but the company won't be selling power. It will be selling confidence, peace of mind, and uninterrupted operations.

That is a model of how to conduct business while adapting to the inverted, connected world. Companies such as Schneider, Hewlett Packard Enterprise, and many others have become acutely aware that as IoT becomes part of more and more lives and businesses, they face a simple reality: adapt or become irrelevant. More to the point, embrace the inversion model of meeting needs instead of pushing product or lose your competitive edge and your market share. Writing in *Business Insider*, technology reporter Matt Weinberger expresses this perfectly:

> Factory owners aren't concerned with servers, storage capacity, or network throughput. They just want to make sure their assembly lines are more efficient, or their hospitals are keeping better track of patients, or their stores have more upsell, whatever that takes.
>
> It means that increasingly, vendors aren't selling a product, or even a set of products.
>
> They're selling efficiency, transparency, and intelligence as a service. A common term for the idea from the IoT Solutions World Congress floor is the idea of charging for "business outcomes," where a customer sets a specific need or goal and the vendor provides whatever technology is necessary to accomplish it.[1]

That's inversion. In the previous chapter, we introduced you to a cross-section of businesses that have adapted their pre-IoT business models to the new connected world order. Regardless of their starting point, these companies understand the dynamic time we are in. As the cost of sensors, actuators, computing, and connectivity continues to fall, and as more businesses and consumers become comfortable, then enthusiastic about the potential of the connected world, the progression from the Internet of Things to the Intelligence of Things to the Immersion of Things will accelerate. Boundaries between products will disappear. Advances such as fog computing, facial recognition, and gestural control will make consuming digitally delivered services as easy and intuitive as opening the morning newspaper.

Most crucial, business-to-business and business-to-consumer constituencies will stop seeing vendors that deliver needs-first solutions via IoT as novel or cool. Instead, they will begin *expecting* the companies that compete for their business to use cloud computing, machine learning, robotics, and predictive analytics, and for more companies to deliver more choice, control, and cost savings. In other words, before long, inversion will simply be the way business is done. Smart business leaders and entrepreneurs are already preparing for that inevitability.

For example, Aviva is a London-based insurance company with more than thirty-three million customers, including millions that purchase auto, truck, and motorcycle insurance. As you know, most, auto insurers offer "good driver" discounts for customers who don't have accidents, but these discounts have always been rather blunt instruments based on customer self-reporting and law enforcement accident records. What if a driver is reckless but lucky? Should someone with poor driving skills who lives in a sparsely populated area receive the same discount as someone who drives in Boston or Paris? With the old system, this was guesswork.

To change this, Aviva launched Aviva Drive, a smartphone app that recruits a phone's accelerometer, orientation sensor, and GPS system to rate a driver's skills in safety-related skills such as braking, cornering, and acceleration. Drive two hundred miles with the app and you get a score; score high enough and you get some substantial safe driving discounts. Aviva is using hardware and software to turn its customers' cars into temporary connected devices—and more importantly, to give customers a way to save money that's based on data and in their control. That is needs-first thinking.[2]

A Culture of Inversion

However, businesses do not take their place at the forefront of the inversion economy simply by accumulating technology, hiring engineers with certain skill sets, or writing new business plans. They do it by building organizations with the capacity to visualize and fully leverage the potential of IoT in all areas, from human resources and product management to IT and executive leadership. They do it by creating a *culture of inversion*.

> *A culture of inversion is an organization-wide environment in which everyone—senior executives, managers, designers, engineers, sales, and so on—understands and supports a business model that begins with "what if" and focuses on meeting needs and creating value and customer experience in new ways, using IoT and other technology as needed.*

It is a culture where the product-first ocean liner is either well into the process of turning (i.e., legacy corporations evolving toward inversion) or has been replaced by a fast, agile, sustainable watercraft (i.e., IoT or sharing economy start-ups). These are the characteristics of a culture of inversion:

- *Needs-first thinking*. The organization's decision makers naturally think in terms of vision and reinventing how their customers interact with a small part of the world. Their concern is not putting product into the supply chain but finding unique, even unprecedented ways to create or add value and enhance the customer experience.
- *Comfort with the technology and architecture of IoT*. At least some of the organization's people are fluent in the language of IoT—sensors, actuators, wireless connectivity, cloud/fog computing, AI, mediation, recruitability, and more. Even those without engineering or computer science backgrounds are conversant in the technology and the inversion vocabulary.
- *A failure-friendly, "launch and learn" mentality*. Leaders of such organizations understand that true innovation often begins with failure, and as a result, they invite risk-taking and discourage the fear of failure. From top to bottom, organizations with a culture of inversion know that human beings truly only believe something is possible after the fact and are eager to champion ideas that sound, on the surface, outlandish or impossible.

- *Openness to collaboration.* Leaders know that as product boundaries disappear, so do boundaries between companies. With the potential of third-party recruitability to add value to connected ecosystems, it will become more important than ever for foresighted organizations to work together.
- *A willingness to play in areas outside historical core competence.* Apple was a computer company, not a phone company. Yet it disrupted and reinvented the mobile phone world. Inversion culture means being open to creating solutions that lie outside your traditional area of expertise but that fit your people's talents and interests.
- *Awareness that the brand value proposition is changing.* In the past, a brand was a promise of the (often sensory and emotional) experience a customer could expect upon buying a product: a sense of "cool," of belonging, of reliability, of comfort. With IoT and the ability to easily update connected devices with new features and track how customers are using solutions, companies are turning experiences into points of tangible value—saving customers' time, increasing convenience, reducing costs, or preventing equipment failures. Value has become less about marketing and more about performance.

Salesforce, a $6.6 billion corporation, probably didn't need to expend resources on IoT to remain relevant in the short term. After all, the San Francisco–based company does make the top-selling customer relationship management solution in the world. However, Salesforce has a culture of inversion. They have estimated that by 2020, there could be seventy-five billion connected devices,[3] and they know that to stay competitive, they need to be part of that ecosystem.

The result: Salesforce Wear, a collection of open-source starter apps that will help customers quickly design and build apps that connect to the Salesforce software platform through wearable devices such as the Myo Gesture Control armband.[4] Most of the characteristics of a culture of inversion—needs-first thinking, collaboration, technical fluency, a willingness to open up the idea of brand value—are present. But what made this venture successful was that Salesforce didn't stop at giving developers a cool set of tools; they factored collaboration into their thinking.

The company provided developers with reference applications showing a wide range of use cases for wearables, allowing developers to get new ideas for how wearable technology could be applied. It provided these reference apps as open-source code, so developers can modify the references and even add their own examples. Finally, Salesforce supported an exhaustive list of wearable devices with the project, increasing the odds that more users will find Salesforce Wear relevant to their needs, increasing adoption—textbook inverted thinking.[5]

The Inversion Triangle

After looking at numerous companies such as Salesforce and other businesses large and small that are driving the development and adoption of IoT, we have identified a mechanism for building an organization driven by a culture of inversion: the *Inversion Triangle*. The model is simple: the three vital components of an inversion organization form the three sides of the triangle, which can be represented as follows:

Technology encompasses the principles of inversion: connectedness, composability, recruitability, immersion, and security/privacy. It's everything that enables an organization to conceive of and develop solutions from a needs-first perspective, from an understanding of IoT architecture and software engineering to expertise in vital IoT skills ranging from circuit design and microcontroller programming to battery life optimization and network security. Technology emerges from the ideas of engineering teams, software developers and coders, industrial designers, and product managers.

Innovation is a product of visionary thinking that should be championed by an organization's chief innovation architect. It's a mentality and organizational structure that encourages leaders to see not only what a company is but what it can evolve into. Innovation thinking empowers and frees designers and engineers to ask "what if" questions and then gives them the resources, time, and support to follow the answers to a new connected device, microchip, sensor, or software application. From an inversion perspective, innovation means having a deep knowledge of what other players in a sector are doing, being

willing to invest in ideas that may have no readily obvious application, and having a plan for initiating and harnessing innovative activity.

Culture is a social, political, and interpersonal fabric within the organization that favors disruption, encourages risk-taking, and delights in delivering new types of value to the customer. It is about leadership that continually recruits the right personnel, rewards bold failures as much as proven winners, fosters open communication and a culture of productive feedback, and discourages siloing of information by encouraging collaboration with other companies. It is also about employees who buy into leadership's vision and are enthusiastic about putting that vision into practice.

In the upcoming section, we will unpack the three pillars of innovation, technology, and culture. However, it is important to understand in each realm, vision is not adequate to produce results. Vision must be successfully executed to achieve inversion. This leads to the ecosystem of inversion, where similar minded companies playing in the inverted world can collaborate effectively.

While it is possible to have a culture of inversion with any two sides of the Inversion Triangle in place, organizations need all three to foster needs-first thinking, create solutions that change markets, and thrive in the connected world. IoT is a technical ecosystem, making engineering virtuosity and fluency in the vocabulary of inversion mission-critical for success. However, the mere presence of technology does not presuppose innovation. Innovative activity and thinking come from a culture that encourages them and become reality when paired with technological resources and know-how.

The Inversion Triangle is a self-reinforcing system that, while it hardly guarantees success in the connected world, certainly increases its likelihood. As we near the conclusion of our exploration of inversion, let's take a closer look at the implications of each side of this triangle for the organizations and leaders who aspire to take their place in this evolving world.

Side One: Technology

In the new economy, we are living not in a proprietary world, but in an *API economy* in which companies that build APIs that connect and enable systems and architectures—that *orchestrate* economic activity—are the new creators of value. Edge computing, machine learning, robotics, and predictive analytics are essential components of the new needs-first economy, but if our goal is to massively, disruptively innovate, then our tools and protocols are only as useful as the thinking that drives them. Today, it is less important to "own" technology than it is to build a modular architecture that you can "use and lose" rather than "develop and safeguard."

While newer inverted companies (which continually experiment with customers even while those customers use their products) are built on that mindset, it is a difficult step change for traditional businesses. For inverted companies, inversion tends to be organic, while for traditional companies, it represents a wrenching ontological shift not only in how they operate but in how they see their place in the world.

In the past, businesspeople were extremely protective of databases and computational models. In fact, until fairly recently, the value proposition with technology appeared to lay in the performance of its components, a proposition not unlike talking about the horsepower or torque of an automobile engine. Computer makers would talk about the clock speeds of their chipsets while Internet service providers bragged about their upload and download speeds, and wireless carriers went on about 3G, then 4G—and now, soon, 5G.

Again, because this is an evolution (the art of re-imagination), not a revolution, those issues still matter, but they are no longer the arbiters of value in technology, and especially in IoT. The performance of components or connectivity in isolation has been replaced by two simple questions, "What can we do with this for our customers?" and "What can our customers do with this for themselves?" Value is no longer about ownership of the technology, but, as we said in chapter 7, about the *orchestration* of all the components of IoT to create solutions that unlock new value for the customer.

Start-ups and young companies have embraced this idea with little friction because it appears a natural outgrowth of the era of Amazon Web Services and software as a service, when former walled gardens such as connectivity and software have not only been turned into services but wildly successful business models. Large legacy corporations, however, often remain trapped in traditional thinking. One example is the fascination with lean/agile methodology. Senior executives at large corporations will say, "I'm applying lean, agile methodology." The problem is, they are applying the lean/agile approach to old technology.

What happens if you put a new, aerodynamic auto body on an old engine and suspension? You have the same car with a face lift. It looks good and gives you the illusion of novelty, but it performs in the same way. The illusion of innovative technology can yield a temporary competitive advantage, but it's an advantage of marketing and brand identity, not substance. In the connected, inverted world, the tipping point with technology is not processing power or bits per second, but *how you think about what technology is and what it can do.* Unless your mindset about technology changes, all you'll have is illusion.

The Experience-Based Economy

To lay a firm foundation for this side of the Inversion Triangle, you must stop thinking about technology as an end in itself—as the delivery system for pixels, packets, bits, photons, money, or anything else, physical or digital. Focusing on such metrics as speed or reliability might seem like common sense, but it's fiercely limiting.

Instead, start thinking of technology—every component, from the sensors, actuators, and microprocessors of an IoT coffeemaker to the engine turbines of a commercial 777—as:

> *A dynamic, interactive platform for curating how your customers experience part of their world.*

In this world, technology is dynamic and interactive because as something powered by software, it can and should always be changing and adapting to the needs of the market. Every time you use a connected product or service, whether you know it or not, data is being captured and analyzed and often, new features are added for real-time testing. Machine learning tests to see whether new things are working, automating a lot of what used to take an entire quality assurance department. Now, with the behavior and rapid feedback of millions of users, the company can perfect what works and quickly eliminate what doesn't. It's business on the balls of your feet, ready to move in any direction. That's agility.

For instance, Uber's SafetyNet function, which can instantly e-mail or text up to five preselected people about your location and estimated time of arrival while you're in transit, is a powerful safety feature. But it wasn't part of Uber on day one; it came about as a result of customer feedback. If Uber had had a rigid architecture and view of its technology, that feature would probably not exist.

Technology is also a curation platform because you can use it to shape how your customers experience a small sliver of their world: agricultural irrigation, finding lodging in a foreign city, locating lost objects, or what have you. Even more powerfully, you can also use it to give them the power to self-select—to *curate*—that experience on their own. What's made Amazon Echo such a seminal product isn't that it lets you order your favorite coffee by shouting across the room; it's that Amazon has surrendered control over how the customer uses Echo to experience their home environment *to the customer*. In other words:

In the product-first model, products became solutions.
In the inverted model, solutions lead to products.

Finally, businesses are reaching the end stage of a transformational process. They began by building products. They inverted their business model and moved to meeting needs. Now they are creating and enabling experiences. We are creating an experience-based economy that goes far beyond collecting and analyzing data and releasing product. The companies featured in chapter 8 are on the vanguard of this, but there are many more.

In this process, we are changing not only the nature of value but the direction from which it flows. Data is evolving and teaching a company's back-end architecture how better to meet needs. In this world, value comes from the place where your solutions and your customers rub together in the real world, where they interact. That is impossible from a product-first, "designers and engineers create the value" mentality. Introducing new features incrementally, testing, and retesting is obsolete.

Orchestration Is Ownership

Grasping this kind of true agility can be extraordinarily difficult for traditional businesses that might have forty, fifty, or even a hundred years of product-first thinking behind them (which is why designating a chief inversion architect can be such an advantage). However, what may even be more challenging is grasping and embracing the changing reality of intellectual property ownership, an idea we touched on earlier.

Until quite recently, the idea of ownership, especially in the competitive tech sector, was ferociously competitive and wrapped up with ego, corporate identity, and culture. Ownership of intellectual property could prove incredibly lucrative, leading to patent trolling, and other unsavory activities. But in the inversion environment, not only is ownership changing, the entire concept of meaningful intellectual property must change as well.

Just as value will no longer flow from how IT components perform but how they create value and allow customers to curate their own experiences, intellectual property will no longer be about the physical device a company has patented. It will be about that company's solution for *orchestrating value*. For example, Strava's business is built on its app-powered, mobile-enabled social network for cyclists and runners. The system collects GPS data from these athletes not only to track metrics such as distance and speed for individual users but to create "heat maps" of cycling activity that Portland, Oregon, and seventy-five other cities are using to see what routes cyclists use, where new bike trails might be needed, where there might be safety concerns, and more.[6] Strava's value does not come from owning bikes, building trails, or designing GPS systems, but from orchestrating activity and data into useful configurations.

This terrifies traditional leaders, who ask, "If we don't own anything, how can we create value for our shareholders?" That is a legitimate concern, so let us address it. What matters is profit, and in the new economy, profit no longer depends exclusively upon ownership of physical or digital intellectual property. In the car world, we are not far from a mainstream experience (and Tesla is there today) where, instead of walking onto a car lot and saying, "I want that blue one with leather interior," you will go to General Motors online (or into a showroom if you crave a more personal or tactile experience) and custom order your car. A display will let you pick everything from your engine and paint color to your navigation and security system, arrange your financing, see when your vehicle will be fabricated and assembled, and schedule shipping to your home. GM will retain and gain market share through its ability to orchestrate the myriad components that go into a modern automobile—and their suppliers—into a single, satisfying experience.

Orchestration is already changing the software world. Though a private company, Airbnb is considered to have a $30 billion valuation—more than Hilton and Hyatt combined.[7] That massive shareholder value doesn't come from owning hotels. Hyatt has huge infrastructure costs: hotels, employees, linens, food, security, you name it. Airbnb owns none of it. It orchestrates. It curates and enhances experiences—and empowers the customer to create and control their own enhanced experiences. Meanwhile, because machine leaning and deep learning run in the background continuously, customers may not realize that their experiences and comments and those of thousands of others are shaping what their future experiences will be.

They created a marketplace to connect the people who have rooms and houses with people who want rooms and houses. The creativity and understanding of technology doesn't rest on speed or power but how they create a frictionless marketplace, maintain quality, ensure security, and give people a tremendous experience. In a world where consumers have the instantaneous ability to express their satisfaction or dissatisfaction, the consumer experience determines the value of technology, not the box or the code.

Side Two: Innovation

We think of innovation as the act of developing new things, but it is actually the ability to think about how to meet a need not by redesigning what already exists but by envisioning what does not yet exist. As we discussed in chapter 7, big companies that still operate as traditional businesses tend to approach innovation in one of two ways:

1. They assume that innovation is only the purview of small, agile start-ups and dismiss it.
2. They try to acquire it.

For example, in March 2016, General Motors spent $1 billion to buy San Francisco self-driving car technology company Cruise Automation.[8] But why? GM has been working on building a driverless car for decades, but the work has remained buried in research. While acquisition is good, GM would put itself ahead of the game if it used its internal resources to fully leverage its acquisition. Similarly, when

Hewlett Packard makes acquisitions that cost millions, odds are they already have all that expertise within their walls, though that expertise may be untapped because the strict corporate culture does not encourage employees to "drive outside their lane." What is the obstacle that keeps these traditional companies from behaving like inverted companies?

One problem: they are too focused on short-term revenue. In this mindset, a market is like a railroad track: it can only proceed in a straight line. It can't go to the left or the right. Many large companies approach innovation from that mindset because everything is about quarterly revenue—about return on investment, recouping the research and development costs of their innovation. In this line of thinking, innovation occurs to serve the product-first model.

As we have seen, this is a recipe for failure. Because innovation is simply a means to an end, the new idea they want to introduce to the market also has to be an instant hit so it can pay for itself—and few innovations actually do that. As a result, ideas die internally because they are not viewed as immediate revenue streams. Inverted companies do not view innovation as the means to make new stuff and create value; they view innovation as the *source* of the value. Innovation is not the means to the end, but the end itself.

When a business is focused on meeting needs first using IoT, software, and wireless connectivity, its ability to innovate, launch and learn, and discard what fails to stick is nearly without limit. Take the mobile app ecosystem, an economy that did not exist before 2007. Now there are more than one million apps on the Apple App Store. As a small app developer, you invest some time creating an app, upload it, and see what happens. If it flops, you learn what you can from user feedback, delete it, and go back to the drawing board. If it's a hit, you have a business. Your sunk cost is practically nothing.

Meanwhile, customers don't pay anything for shipping or spend time shopping. The environment is so frictionless for users that they're willing to try new apps, keep what works, and discard the rest. Ownership isn't an issue. What delivers a great experience, what helps them curate their world a little better—that's what creates value.

This makes continuous innovation not only possible but essential. In the inverted world, companies can no longer think of innovation as a buzzword or even as a laboratory for gestating new things to slip into the product-first supply chain. Innovation must become your company's operating system. Inverted companies lead with innovation—they start by asking "What if?"—and then do it continuously, relying on the interactive flow of use and behavior data powered by IoT.

Innovating Experiences

The other challenging idea for traditional businesses is that *what* they are innovating has changed. A few years ago, Netflix was that company that diligently mailed you movies on DVD. It was a nice business model, nice enough to take down Blockbuster and its nationwide network of stores. But it was finite. There were only so many DVDs, so you would get in a queue.

Then came Redbox. Now you had a machine that sat in your grocery store where you could walk up and get DVDs when you wanted them. It was a direct challenge to Netflix because you no longer had to wait by your mailbox. And if Netflix had thought like a traditional business and said, "We're in the business of renting DVDs," and tried to innovate around the business of sending physical products through the mail, Redbox might have eventually forced them out of business.

But Netflix didn't do that. They said, "We're an entertainment company," and got in on the beginning of the streaming movie market. They innovated, and after they did it once, they kept doing it. Streaming movies evolved into making new movies and television series. Think about that level of agility: to go from being Blockbuster in the mail to the young company that is not only disrupting video rental but disrupting Hollywood, content delivery, and the whole experience model.

These examples point to one more key to companies following the inversion model becoming successful innovators. If you want to innovate and build something extraordinary, put real resources behind your innovators. Don't get into a market halfway. Invest financially and psychologically in making your innovation work. Incrementalism will ruin you. For example, Netflix pays its engineers extremely well and gives them some of the longest maternity and paternity leaves in the tech world. As a result, it has the best and the brightest, and it needs them.

Why? Because a few years ago, when Netflix began making its own programming such as *House of Cards*, they experienced what could be described as a classic unintended consequence: *binge watching*. Five million people would watch the entire season of a show at once. What if Netflix had "sort of" gone into streaming, and then panicked: "Oh my God, we have five million people streaming the entire season of *House of Cards*. What do we do?" They probably would have pulled the plug and someone else would have been first. But by having a needs-first, inversion mindset and thinking about innovation as something central, not optional, Netflix created "binge culture," which has a direct impact on revenue because millions more people experienced millions of additional hours of content. *That is innovation-induced behavior.*

That's what the inversion model allows. It's about using connected technology to meet needs in new ways, so you release, continuously analyze, and if something works, you build on it. If it doesn't, you let it go. Had Netflix been a traditional company, they would have gone down with the old model and nothing would have changed. Instead, Netflix not only succeeded in streaming but came to dominate the original programming space and inspire other players such as Amazon and Hulu to create their own original content as well. How far has this disruption progressed? In 2016, *Manchester by the Sea*—a film released and distributed by Amazon Studios—won two Academy Awards.

The mobile interface has heightened expectations even more. When you deliver an experience through the mobile interface, you have a connection to that customer. Another excellent example is Pandora. Some people think Pandora is about music, but it's really about creating a personal soundscape, something no one else has ever done. Pandora knows what music you love and dislike and can create experiences for you based on those tastes. When an artist is in town, they can even send you an alert that says, "The Lumineers are in town July 10. Tickets are still available." Again, you're curating and innovating experiences.

Side Three: Culture

That leads us to culture, the internal social, political, and institutional environment of an organization. You can embrace the inversion model and have the technology and innovation mindset to support the venture you're going to take. But that's not sufficient unless you have a culture that supports risk and takes delight in disruption.

First, culture is not your mission statement. Some companies will proudly share their mission statements and might even include vague language about innovation in them. Then you look at their history and think, "What have you introduced to the market that's new in the last 20 years?"

There are two kinds of culture in the inverted world. One is top-down culture, which is not just a matter of leadership. Inversion requires not just superb leadership, but *visionary* leadership. Traditional businesses bring in people who know how to run the business, scale the business, build partnerships, and maybe even open offices all over the world. But that's not sufficient for inversion. A company moving toward an inversion model needs a visionary leader who can see untapped markets and unrealized value. In this economy, successful companies don't just have business leaders who know how to manage. They have visionaries who imagine markets that don't yet exist and lead their people to use technology to create those markets.

> *When you're a visionary leader, you're constantly pushing the "what if" button, not the revenue button. Revenue is a consequence of vision.*

When you pursue revenue, your main concern becomes maintaining that status quo. If Netflix had been run by a traditional CEO, they would have never ventured into streaming. They would have said, "Our whole business model is shipping DVDs. It's working fine and we're profitable. Imagine if we take this risk and it fails."

That sort of visionary leadership requires courage, fearlessness, and a deep knowledge of what the future can be even when it isn't apparent. Most of all, it demands conviction that starting with needs is the only path forward. Steve Jobs already had a wildly successful computer company, yet in the space of nine years, he went from changing music to obliterating the old telco business model to creating the tablet market—and creating the app economy to boot. A few years before, none of this was there. He and a few others saw it and made it happen.

As we have said, you don't need to be Jobs to practice inversion, but you do have to start. After all, Jobs didn't know he would become Jobs when he started, either. *Every traditional company has what it takes to invert and move forcefully into inversion.* There are no limitations in resources or resilience. Only in the courage to lead with vision.

An Unlikely Culture of Inversion

If you doubt this, consider Microsoft's stunning evolution. Its resurgence suggests that a culture of inversion is not only the province of the small, agile start-up. Even the largest established corporation can evolve into a needs-first organization if leadership exhibits the necessary vision and determination.

As recently as the fall of 2014, analysts were calling Microsoft a dinosaur.[9] Microsoft was clinging to its legacy business of selling Office productivity suites, had made an ill-advised acquisition of Nokia, and Windows 8 had been poorly received. But then along came CEO Satya Nadella, and the ship began to turn. The company began making bold bets and winning many of them: the HoloLens "mixed reality" headset; the two-device-in-one Surface, which was initially panned but has become a sensation not only as a computer but a soon-to-be-released mobile phone; Windows 10, which landed to rave reviews. Most importantly, Microsoft pivoted hard to the cloud with the Azure platform, bringing to the market not merely cloud computing but cloud-based AI.

The result? Microsoft is a cool company again, with a rising stock price and the sense that the company is once again a hot landing spot for top IT talent. To be sure, the Redmond titan is far from done with its inversion journey, and it remains a product-driven company in many ways, but what Nadella and his team have accomplished in such a short time is remarkable.

Not only did the company completely change its product development path, making all its existing products cloud-enabled, but it radically changed the corporate culture. Not long ago, Microsoft was the epitome of the hidebound, change-resistant giant infected with frame inertia. No more. When they pivoted, they pivoted in their culture, not just their technology. One of Nadella's key moves in enabling this shift was to embrace the "growth mindset," an idea at the core of the 2006 book *Mindset: The New Psychology of Success*, by Stanford professor of Psychology Carol Dweck.[10] The book has become required reading at Microsoft, and one of its core concepts underpins the company's new approach to optimizing its employees' effectiveness.

Now, instead of assuming that individuals' skills are fixed, the evolving Microsoft culture recognizes that success is often the result of hard work and discovery, not innate talent. The company's leaders look for opportunities to pursue improvement not only through learning from others but also by learning from one's own mistakes. The goal: discover what you don't know, learn, move quickly, and identify the right path forward.

This radical shift is reflected in a now-famous internal 2015 e-mail from Nadella published by *GeekWire*. In part, it reads:

> We need to be always learning and insatiably curious. We need to be willing to lean in to uncertainty, take risks and move quickly when we make mistakes, recognizing failure happens along the way to mastery. And we need to be open to the ideas of others, where the success of others does not diminish our own. . . . We will learn about our customers and their businesses with a beginner's mind and then bring solutions that meet their needs. We will be insatiable in our desire to learn from the outside and bring that knowledge into Microsoft, while still innovating to surprise and delight our users.[11]

Microsoft is not yet a pure needs-first business and may never be fully so. But its evolution demonstrates that the successful adoption of inversion principles and an inversion mindset is within reach of any company, of any size.

Culture Is the New Brand

Today, customers in the business-to-business and business-to-consumer worlds know more about the companies that meet their needs than ever. Because of this, the culture of the entire organization telegraphs as much of its promise as its solutions and its marketing. In a transparent, connected world, culture has become brand. But this culture is bottom-up. It encourages people to think in "what if" terms and doesn't try to herd them toward a single objective. It's no longer sufficient to say, "Operate in an agile model but do exactly as I say."

In an inverted company, the irreplaceable bottom-up cultural pieces are *freedom* and *creativity*. Inversion culture is creative, not prescriptive. People insist on freedom to think, create, and curate their own experience just like you're trying to let customers curate theirs. People aren't stuck in one zone, one sector of work. They can stretch into new zones and get excited about the possibilities. Innovation is a natural outcome of this culture; that's one of the factors that makes Amazon so successful. Every time you visit Amazon.com, there's a good chance you will see something extraordinary.

Does that mean everyone at Amazon is creative? Of course not. But the culture of the company is comfortable innovating at high speeds to enhance customer experience. The energy of such companies carries forward to the customer experience. There's an internal "what do we get to create next?" excitement that breeds a "what are they going to do next?" anticipation and admiration in customers. That culture not only impacts what the company produces, but also what the company becomes.

Because of the way it interfaces with the consumer, inversion culture is also highly transparent. People like Tesla's culture because their expectation of Tesla is very different from any other car company, starting with how they interact in buying a car. They're not walking on a snow-covered car lot with a chattering, pushy salesman. They're having a self-curated experience. The culture of Tesla is that everything is exposed. There are no secrets, because they know that today's customer has near-infinite choice and near-infinite expectations. Aspects of companies such as Tesla that were once kept under wraps are now assets to be used, criticized, measured, and leveraged.

Culture and product are no longer separate. The external culture of the customer is also starting to blur with internal culture of the employee to create a shared experience.

> *The logical conclusion is that in this new connected world, culture is brand.*

Customers are able to know the people in a company (aided by social media), its internal operations, and its mindset, and they get a strong sense of what that company will be like to work with and whether it will deliver on its promises. Conventional wisdom has always said that culture is everything about a company, and that idea remains true—but in the inversion era, it's the company itself that is changing.

Inversion is best considered as a triangle because all three sides are essential, none more important than another. Also, all three sides are the same length, indicating that the transition to inverted forms of technology, innovation, and culture must happen simultaneously. You can't pursue innovative vision while your culture lags in the traditional dust or invest in the technology of IoT without the innovative thinking to use it. In the end, it is the harmonious balance of these three elements—technology, innovation, and culture—working together in a constantly evolving world where global is now local, that turn traditional dinosaurs into inverted high flyers.

Are You Ready?

After all this, you may feel a bit disrupted and unsettled. Fantastic. That is normal and healthy, but the challenge is, where do you go from here? We hope we have made the path, and the journey, clear. There are thousands of examples of companies that have enjoyed great success in the inverted, connected world—some of them starting with traditional limitations and then transcending them. We have broken down inversion—shown you the source code, in a way—so that you can use this to begin your organization's transformation—or continue it.

The Conversation Continues

While writing this book, we've been building a digital platform where the exchange of ideas will go on. We encourage you to participate with us, to share your ideas and brilliant innovations by visiting inversionFactor.com.

Now, go and ask, "What if?"

You have come a long way and are to be congratulated. You have read the story of inversion, learned its principles, explored the evolution of things in IoT, discovered the keys to becoming an inverted organization, and read dynamic examples of how such organizations have made inversion work.

Together we have discussed the role of technology as a platform, talked about innovation as an operating system, and shown how culture ties everything together. But you need all three sides of the Inversion Triangle to reach the success at its center. You need innovative thinking to see your company as something new. You need technical expertise to build a solution. And you need culture to create the environment in which both can occur. Technology is constantly changing. Innovation and customer experiences are fluid. Culture is as dynamic as the people who make it.

The next step is yours.

Are you ready?

Notes

Introduction

1. John McElroy, "The Race for Autonomous Cars Is Over: Silicon Valley Lost," *Auto Blog*, February 17, 2017, www.autoblog.com/2017/02/21/race-for-autonomous-cars-is-over-mcelroy-autoline-opinion/.

2. Gartner, "Gartner Says 6.4 Billion Connected 'Things' Will Be in Use in 2016, Up 30 Percent from 2015," news release, November 10, 2015, http://www.gartner.com/newsroom/id/3165317.

3. James Macaulay, Lauren Buckalew, Gina Chung, *Internet of Things in Logistics* (Troisdorf, Germany: DHL Trend Research and Cisco Consulting Services, 2015).

4. James Manyika, Michael Chui, Peter Bisson, Jonathan Woetzel, Richard Dobbs, Jacques Bughin, and Dan Aharon, *The Internet of Things: Mapping the Value beyond the Hype* (New York: McKinsey Global Institute, 2015).

5. It is important to recognize that the "what if" economy must have strong boundaries regarding ethics, privacy protection, and organizational and individual behavior. You don't "what if" following the law or acting fairly and ethically.

6. "BACCH™ 3D Sound: A Revolutionary Technology for Audiophile-Grade 3D Audio," 3D Audio & Applied Acoustics Lab, Princeton University, accessed July 17, 2017, https://www.princeton.edu/3D3A/PureStereo/Pure_Stereo.html.

7. This concept shares DNA with two others: *design thinking* and Harvard Business School professor Theodore Levitt's famous statement, "People don't want to buy a quarter-inch drill. They want a quarter-inch hole!" Are you in the business of selling drills or holes? We would argue that IKEA's DIY furniture is actually an example of selling holes. The holes come predrilled, so they actually compete with Black & Decker. That is an early example of inversion.

8. Ethics and historical accuracy compel us to acknowledge the claim that engineer and entrepreneur Peter T. Lewis first coined, spoke about, and published the term "Internet of Things" in a speech to the Congressional Black Caucus Foundation 15th Annual Legislative Weekend in Washington, D.C. in September 1985.

Chetan Sharma, "Correcting the IoT History," Chetan Sharma Consulting, accessed July 17, 2017, http://www.chetansharma.com/correcting-the-iot-history/.

Chapter 1

1. V2G, or Vehicle to Grid, is a concept that has been studied widely. The vehicle communicates with the electric grid and either sells power to the grid or reduces consumption to improve the performance of the grid. The power can also run your home when you have an outage, for example. Jim Motavalli, "Power to the People: Run Your House on a Prius," *New York Times*, September 2, 2007, http://www.nytimes.com/2007/09/02/automobiles/02POWER.html.

2. Self-driving technology is becoming commoditized, but vehicles can gain further benefits from driving in a train, bumper-to-bumper at high speed. This is called platooning. It reduces drag, much like cyclists in a peloton; reduces the chances of collision; and increases the capacities of existing roads.

Ann Hsu, Farokh Eskafi, Sonia Sachs, and Pravin Varaiya, *Design of Platoon Maneuver Protocols for IVHS* (Berkeley: California Partners for Advanced Transit and Highways [PATH], 1991); Pooja Kavathekar and YangQuan Chen, "Vehicle Platooning: A Brief Survey and Categorization" (presentation, Proceedings of the ASME 2011 International Design Engineering Technical Conferences and Computers and Information in Engineering Conference IDETC/CIE 2011, Washington, DC, August 28–31, 2011).

3. Remi Tachet, Paolo Santi, Stanislav Sobolevsky, Luis Ignacio Reyes-Castro, Emilio Frazzoli, Dirk Helbing, and Carlo Ratti, "Revisiting Street Intersections Using Slot-Based Systems," *PloS One* 11, no. 3 (2016): e0149607.

4. Cruising for parking can contribute significantly to traffic congestion: from 8 to 70 percent depending on the situation. Donald C. Shoup, "Cruising for Parking," *Transport Policy* 13, no. 6 (2006): 479–486.

5. Advances in battery technology, though significant, do not follow Moore's Law, which is exponential. Charging rates and the number of charges will remain limited. Monitoring the state of charge on the battery will be necessary to ensure performance. Furthermore, when charging requires a hundred kilowatts per

vehicle, the grids need to compensate. Advance information and scheduling will help balance the loads.

6. Home delivery robots are not surprisingly close. Companies such as Starship and Piaggio Fast Forward are already showing products. "The Future of Home Delivery: Pedestrians and Robots Will Soon Share the Pavements," *Economist*, February 18, 2017, http://www.economist.com/news/science-and-technology/21717025-streetwalkers-pedestrians-and-robots-will-soon-share-pavements.

7. Technologies such as IoT, self-driving cars, and electric vehicles will help in managing climate change by optimizing resource use and by spreading the embodied costs of manufacturing across multiple functions. Robert Costanza, "Embodied Energy and Economic Valuation," *Science* 210, no. 4475 (1980): 1219–1224.

8. Virtual assistants such as x.ai are already products, albeit in early stages of functionality. Over time, interaction with IoT systems will flourish.

9. Tadas Baltrušaitis, Daniel McDuff, Ntombikayise Banda, Marwa Mahmoud, Rana el Kaliouby, Peter Robinson, and Rosalind Picard, "Real-Time Inference of Mental States from Facial Expressions and Upper Body Gestures" (presentation, Automatic Face & Gesture Recognition and Workshops [FG 2011], 2011 IEEE International Conference on Pervasive Computing and Communications Workshops, Santa Barbara, California, March 21–25, 2011).

10. Music does reduce stress—if it is classical. Elise Labbé, Nicholas Schmidt, Jonathan Babin, and Martha Pharr, "Coping with Stress: The Effectiveness of Different Types of Music," *Applied Psychophysiology and Biofeedback* 32, nos. 3–4 (2007): 163–168.

11. Gartner, "Gartner Survey Shows That 43 Percent of Organizations Are Using or Plan to Implement the Internet of Things in 2016," news release, March, 3, 2016, http://www.gartner.com/newsroom/id/3236718.

12. We discuss Uber on numerous occasions because they have engaged in a significant number of bold, successful ventures that are excellent examples of the power of inverted, needs-first thinking. However, this should not be construed as an endorsement of Uber's corporate culture or the actions and attitudes of its personnel.

13. "Motorola Ships 50 Millionth MotoRazr," news release, July 18, 2006, http://www.upi.com/Motorola-ships-50-millionth-RAZR-V3/78181153240991/.

14. Eric Ries, *The Lean Startup: How Today's Entrepreneurs Use Continuous Innovation to Create Radically Successful Businesses* (New York: Crown Business, 2011).

Chapter 2

1. Katie Fehrenbacher, "The Zipcar for Electric Scooters Grows Up and Out," *Fortune*, July 9, 2015, http://fortune.com/2015/07/09/zipcar-for-electric-scooters/.

2. Scoot Networks, "Urban Transit Pioneer, Scoot Networks, Unveils Blueprint for Shared Electric Mobility to Solve Traffic Congestion and Emission Challenges in World's Largest Cities," press release, StreetInsider.com, March 9, 2017, www.streetinsider.com/Press+Releases/Urban+Transit+Pioneer%2C+Scoot+Networks%2C+Unveils+Blueprint+for+Shared+Electric+Mobility+to+Solve+Traffic+Congestion+and+Emission+Challenges+in+World%27s+Largest+Cities/12649475.html.

3. Globe Newswire, "Sensoria and VIVOBAREFOOT Showcase First Internet Connected Running Shoe Designed for Natural Running," news release, Nasdaq Globe Newswire, January 4, 2017, https://globenewswire.com/news-release/2017/01/04/903234/0/en/Sensoria-and-VIVOBAREFOOT-Showcase-First-Internet-Connected-Running-Shoe-Designed-for-Natural-Running.html.

4. James Manyika, Michael Chui, Peter Bisson, Jonathan Woetzel, Richard Dobbs, Jacques Bughin, and Dan Aharon, *The Internet of Things: Mapping the Value beyond the Hype* (New York: McKinsey Global Institute, 2015).

5. *IoT Case Studies: Companies Leading the Connected Economy* (New York: American International Group, 2016).

6. Derek du Preez, "ServiceMax—Turning Manufacturing Companies into IoT Enabled Utilities," *diginomica*, September 26, 2016, http://diginomica.com/2016/09/26/servicemax-turning-manufacturing-companies-into-iot-enabled-utilities/.

Chapter 3

1. See the PATH website at http://www.path.berkeley.edu.

2. Nara Shin, "Raden Smart Luggage," *Cool Hunting*, March 29, 2016, http://www.coolhunting.com/travel/raden-smart-luggage-travel-tech.

3. Nicky Woolf, "DDoS Attack That Disrupted Internet Was Largest of Its Kind in History, Experts Say," *Guardian*, October 26, 2016, https://www.theguardian.com/technology/2016/oct/26/ddos-attack-dyn-mirai-botnet.

4. Common Criteria, "Common Criteria for Information Technology Security Evaluation, Part 1: Introduction and General Model," September 2012, https://www.commoncriteriaportal.org/files/ccfiles/CCPART1V3.1R4.pdf.

Chapter 4

1. Organization for Economic Co-Operation and Development, *The Future of the Internet Economy: A Statistical Profile* (Paris: OECD, 2011).

2. Krishna M. Kavi, "Beyond the Black Box," *IEEE Spectrum* 47, no. 8 (2010): 46–51.

3. Rolls Royce, "Engine Health Management," Rolls Royce corporate website, accessed February 22, 2017, https://www.rolls-royce.com/about/our-technology/enabling-technologies/engine-health-management.aspx#analyse.

4. For more information, see "Water Use Efficiency (in Cities): Leakage" European Environment Agency, http://www.eea.europa.eu/data-and-maps/indicators/water-use-efficiency-in-cities-leakage.

5. We must acknowledge that in its race to disrupt the home security industry, SimpliSafe also allowed flaws to creep into its products that left its customers vulnerable to beginner hackers, highlighting a vital truth of inversion: an inverted solution must be needs-based, but it must also be superior to the old solution. See Thomas Fox-Brewster, "300,000 American Homes Open to Hacks of 'Unfixable' SimpliSafe Alarm," *Forbes*, February 17, 2016, https://www.forbes.com/sites/thomasbrewster/2016/02/17/simplisafe-alarm-attacks/#325966ac3b00.

6. Kyle Campbell, "U.S. Regulators to Close Tesla Autopilot Crash Investigation, No Recall Expected," *New York Daily News*, January 19, 2017, http://www.nydailynews.com/autos/news/no-recall-expected-tesla-autopilot-probe-article-1.2950333.

Chapter 5

1. Fabio Vergi, "The Evolution of Mobile Phones (Infographic)," *Let's Talk Tech* (blog), September 18, 2015, http://www.letstalk-tech.com/the-evolution-of-mobile-phones-infographic/.

2. Marc Andreessen, "Why Software Is Eating the World," *Wall Street Journal*, August 20, 2011, http://a16z.com/2016/08/20/why-software-is-eating-the-world/.

3. Alex Brisbourne, "Tesla's Over-the-Air Fix: Best Example Yet of the Internet of Things?," *Wired*, February 2014, https://www.wired.com/insights/2014/02/teslas-air-fix-best-example-yet-internet-things/.

4. Matt Novak, "Hackers Shut Down the Key Card Machine in This Hotel until a Bitcoin Ransom Was Paid [Corrected]," *Gizmodo* (blog), January 30, 2017,

http://gizmodo.com/hackers-locked-every-room-in-this-hotel-until-a-bitcoin-1791769502.

5. Mark Harris, "Researcher Hacks Self-Driving Car Sensors," *IEEE Spectrum*, September 4, 2015, http://spectrum.ieee.org/cars-that-think/transportation/self-driving/researcher-hacks-selfdriving-car-sensors.

Chapter 6

1. *Industrie 4.0: Innovationen für die Produktion von Morgen* (Bonn, Germany: Bundesministerium für Bildung und Forschung, 2016).

2. Richard Gendal Brown, "Introducing R3 Corda: A Distributed Ledger Designed for Financial Services," *Richard Gendal Brown* (blog), April 5, 2016, gendal.me/2016/04/05/introducing-r3-corda-a-distributed-ledger-designed-for-financial-services/.

3. Theodora Koullias, an MIT alumna, has been developing a smart handbag and described the basic concept to Sanjay in 2016. Information can be found at http://www.jonlou.com.

4. KGS Buildings, "Proactive Campus-Wide Maintenance Results in Significant Energy Reductions," case study, KGS Buildings, https://cdn2.hubspot.net/hubfs/612214/KGS_Buildings_Education_Case_Study.pdf.

5. Brian J. Dlouhy, Brian K. Gehlbach, Collin J. Kreple, Hiroto Kawasaki, Hiroyuki Oya, Colin Buzza, and Mark A. Granner et al., "Breathing Inhibited When Seizures Spread to the Amygdala and upon Amygdala Stimulation," *Journal of Neuroscience 35*, note 28 (2015): 10281–10289.

6. Rob Matheson, "Wristband Detects and Alerts for Seizures, Monitors Stress," *MIT News*, March 9, 2016, http://news.mit.edu/2016/empatica-wristband-detects-alerts-seizures-monitors-stress-0309.

7. Miles Johnson, "Hyundai Adds Blue Link Skill for Amazon Alexa," press release, November 15, 2016, http://www.hyundainews.com/us/en/media/pressreleases/46590/hyundai-adds-blue-link-skill-for-amazon-alexa.

8. DHL, "DHL Now Delivers Parcels to Smart Car Trunks," press release, July 25, 2016, http://www.dhl.com/en/press/releases/releases_2016/all/parcel_ecommerce/dhl_now_delivers_parcels_to_smart_car_trunks.html.

9. Lucian Constantin, "New Insulin Pump Flaws Highlight Security Risks from Medical Devices," *Computerworld*, October 4, 2016, http://www.computerworld.com/article/3127148/security/new-insulin-pump-flaws-highlight-security-risks-from-medical-devices.html.

10. Kimiko de Freytas-Tamura, "The Bright-Eyed Talking Doll That Just Might Be a Spy," *New York Times*, February 17, 2017, https://www.nytimes.com/2017/02/17/technology/cayla-talking-doll-hackers.html.

11. Kim Zetter, "Inside the Cunning, Unprecedented Hack of Ukraine's Power Grid," *Wired*, March 3, 2016, https://www.wired.com/2016/03/inside-cunning-unprecedented-hack-ukraines-power-grid/.

12. Yossi Melman, "Computer Virus in Iran Actually Targeted Larger Nuclear Facility," *Haaretz*, September 28, 2010, http://www.haaretz.com/computer-virus-in-iran-actually-targeted-larger-nuclear-facility-1.316052.

13. Aaron Tilley, "How a Few Words to Apple's Siri Unlocked a Man's Front Door," *Forbes*, September 21, 2016, http://www.forbes.com/sites/aarontilley/2016/09/21/apple-homekit-siri-security/#586ff9c86e8a.

14. Joshua Siegel, "Data Proxies, the Cognitive Layer, and Application Locality: Enablers of Cloud-Connected Vehicles and Next-Generation Internet of Things" (PhD dissertation, Massachusetts Institute of Technology, 2016).

Chapter 7

1. David Floyd, "Who Killed Sears? 50 Years on the Road to Ruin," Investopedia, February 16, 2017, http://www.investopedia.com/news/downfall-of-sears.

2. Niam Yaraghi and Shamika Ravi, "The Current and Future State of the Sharing Economy," Brookings, December 29, 2016, https://www.brookings.edu/research/the-current-and-future-state-of-the-sharing-economy/.

3. Matthew L. Wald, "Smart Electric Utility Meters, Intended to Create Savings, Instead Prompt Revolt," *New York Times*, December 13, 2009, www.nytimes.com/2009/12/14/us/14meters.html.

4. EIA, "How Many Smart Meters Are Installed in the United States, and Who Has Them?," U.S. Energy Information Administration, December 7, 2016, www.eia.gov/tools/faqs/faq.php?id=108&t=3.

5. Jonathan Shieber, "Roadie Launches as the Uber for Shipping and Delivery," *TechCrunch*, January 27, 2015, https://techcrunch.com/2015/01/27/roadie-launches-as-the-uber-for-shipping-and-delivery/.

6. AeroMobil, "World Premiere: Aeromobil Will Announce Its Next Generation Flying Car at Top Marques Monaco 2017," news release, April 12, 2017, https://www.aeromobil.com/official-news/detail/world-premiere-aeromobil-will-announce-its-next-generation-flying-car-at-top-marques-monaco-2017/.

7. Phys.org, "Dubai aims to launch hover-taxi by July," news release, February 13, 2017, https://phys.org/news/2017-02-passenger-carrying-drone-dubai.html.

8. Akshat Rathi, "IBM Researchers Have Created an 'Impossible' Molecule That Could Power Quantum Computers," *Quartz*, February 14, 2017, https://qz.com/910146/ibm-ibm-researchers-have-created-a-triangular-molecule-that-chemists-thought-was-impossible-and-it-could-power-quantum-computers/.

9. Andrew Krok, "Volkswagen Is Taking on Uber with Moia, Its New Ride-Hailing Startup," *Road Show*, December 5, 2016, https://www.cnet.com/roadshow/news/volkswagen-to-face-sharing-economy-titans-with-new-brand-moia/.

10. Linda Bernardi, *Provoke: Why the Global Culture of Disruption is the Only Hope for Innovation* (Seattle: TerraVanDeVitis, 2009), 24.

Chapter 9

1. Matt Weinberger, "This Is How the 'Internet of Things' Is Changing the Business Model of the World's Biggest Technology Companies," *Business Insider*, October 27, 2016, http://www.businessinsider.com/hewlett-packard-enterprise-cisco-schneider-electric-iot-2016-10.

2. Aviva, "Aviva Launches Its New Aviva Drive App," news release, November 26, 2012, https://www.aviva.co.uk/media-centre/story/17048/aviva-launches-its-new-aviva-drive-app/.

3. Stuart Leung, "5 Ways the Internet of Things Will Make Marketing Smarter," *Salesforce Blog*, March 20, 2014, https://www.salesforce.com/blog/2014/03/internet-of-things-marketing-impact.html.

4. Salesforce.com, "Salesforce.com Launches Salesforce Wear," news release, June 10, 2014, https://www.salesforce.com/company/news-press/press-releases/2014/06/140611.jsp.

5. Denis Pombriant, "Salesforce Wear, What Does That Mean?," *Enterprise Irregulars*, June 17, 2014, www.enterpriseirregulars.com/75368/salesforce-wear-mean/.

6. Peter Walker, "City Planners Tap into Wealth of Cycling Data from Strava Tracking App," *Guardian*, May 9, 2016, https://www.theguardian.com/lifeandstyle/2016/may/09/city-planners-cycling-data-strava-tracking-app.

7. Deanna Ting, "Airbnb's Latest Investment Values It as Much as Hilton and Hyatt Combined," Skift, September 23, 2016, https://skift.com/2016/09/23/airbnbs-latest-investment-values-it-as-much-as-hilton-and-hyatt-combined/.

8. Mike Colias, "GM to Buy Autonomous-Driving Startup Cruise," *Automotive News*, March 11, 2016, http://www.autonews.com/article/20160311/OEM06/160319974/gm-to-buy-autonomous-driving-startup-cruise.

9. Michael Ide, "Microsoft Corporation Is a Digital Dinosaur, Says Indigo," *ValueWalk*, September 2, 2014, www.valuewalk.com/2014/09/microsoft-digital-dinosaur-says-indigo.

10. Matt Weinberger, "Satya Nadella Says This Book Gave Him the 'Intuition' He Needed to Revamp Microsoft," *Business Insider*, August 4, 2016, www.businessinsider.com/microsoft-ceo-satya-nadella-on-growth-mindset-2016-8.

11. Todd Bishop, "Exclusive: Satya Nadella Reveals Microsoft's New Mission Statement, Sees 'Tough Choices' Ahead," *GeekWire*, June 25, 2015, www.geekwire.com/2015/exclusive-satya-nadella-reveals-microsofts-new-mission-statement-sees-more-tough-choices-ahead/.

Index